STUDY GUIDE

GENERAL CHEMISTRY
Principles and Structure
Third Edition

JAMES E. BRADY
St. John's University
New York

JOHN WILEY & SONS
New York Chichester Brisbane Toronto Singapore

PREFACE

In this third edition I have retained the same overall format of the Study Guide because so many students have found it useful. Because students usually study a chapter one section at a time, the Study Guide is divided into sections that exactly parallel those in the textbook. Each begins with a statement of objectives that prepares the student for what is to be presented and outlines what he or she is to accomplish. Also included in each section is a brief review. Here the key concepts that are developed in the section are summarized to bring them into focus. In some instances, especially in the critical early chapters, additional explanations of major topics are presented. In this edition I have added additional worked-out examples where appropriate to further complement those that appear in the textbook.

Most sections contain a brief self-test intended to allow students to test their mastery of the subject matter before moving on. Many of these self-tests have been expanded. They generally provide questions of graded difficulty to permit students to progress from simple problems in the direction of more complex ones. None of the problems are very difficult, however, since there is an ample number of difficult problems (those marked with an asterisk) available in the text itself.

Each section concludes with a list of new terms that have appeared in the corresponding section of the text. Students are urged to learn the meanings of these new terms before moving on.

In this edition the detailed solutions to selected numerical problems have been omitted because they and others are now available in a separate supplement. In their place is a Glossary that provides definitions of all the "new terms" that are listed in the Study Guide. Placement of the Glossary in the Study Guide was chosen so that students could easily refer to it, without losing their place, while studying chapters in the textbook.

James E. Brady

CONTENTS

BEFORE YOU BEGIN...

Before you begin your general chemistry course, read the next several pages. They're designed to tell you how to use this study guide and to give you a few tips on improving your study habits.

How to Use the Study Guide

This book has been written to parallel the topics covered in your text, General Chemistry: Principles and Structure. For each section in the text you'll find a corresponding section in the study guide. In the study guide the sections are divided into Objectives, Review, Self-Test and New Terms. Before you read a section in the text, read the Objectives in the study guide. This will give you a feeling for what to keep an eye on as you read the text. It should help you understand what you must pay attention to.

After you've read a section, return to the study guide and read the Review. This will point out specific ideas that you should be sure you have learned. Sometimes you will be referred back to the text to review specific items there. Sometimes there will be additional explanations of difficult or important concepts, and in some instances there will be additional worked-out sample problems. Work with the review and the text together to be sure you have mastered the material before going on.

In most sections you will also find a short Self-Test to enable you to test your knowledge and problem-solving ability. The answers to all of the self-test questions are located at the end of the chapters in the study guide. Try to answer the questions without having to look at the answers. A space is left after each question so that you can write in your answers and then check them all after you've finished.

An important aspect of learning chemistry is becoming familiar with the language. There are many cases where a lack of understanding can be traced to a lack of familiarity with some of the terms used in a discussion or a problem. A great deal of effort was made in your textbook to avoid using a term without first adequately defining it. Once it has been defined, however, it's used with the assumption that you've learned its meaning. It's important, therefore, to learn new terms as they appear, and for that reason, most of them are set in boldface type in the text. At the end of each section of the study guide there is a list of New Terms. Look them over and check to be sure you know their meaning before proceeding on to the next section. You might find it worthwhile to write out their meanings in your chemistry notebook. Then check yourself by referring to the Glossary at the back of the study guide. This will help you review important terms later when you prepare for quizzes or examinations.

Study Habits

You say you want to get an A in chemistry? That's not as impossible as you may have been led to believe, but it is going to take some work. The key is efficient study, so your precious study time isn't wasted. Efficient study requires a regular routine, not hard study one night and nothing the next. At first it is difficult to train yourself, but after a short time your study routine will indeed become a study habit and your chances of success in chemistry, or any other subject, will be greatly improved.

To help you get more out of class, try to devote a few minutes the evening before to reading, in the text, the topics that you will cover the next day. Read the material quickly just to get a feel for what the topics are about. Don't worry if you don't understand everything; the idea at this stage is to be aware of what your teacher will be talking about.

Your lecture instructor and your textbook serve to complement one another; they provide you with two views of the same subject. Try to attend lecture regularly and take notes during class. These should include not only those things your teacher writes on the blackboard, but also the important points he or she makes verbally. If you pay attention carefully to what your teacher is saying in class, your notes will probably be somewhat sketchy. They should, however, give an indication cf the major ideas. After class, when you have a few minutes, look over your notes and try to fill in the bare spots while the lecture is still fresh in your mind. This will save you much time later when you finally get around to studying your notes in detail.

In the evening (or whatever part of the day you close yourself off from the rest of the world to really study intensely) review your class notes once again. Use the text and study guide as directed above and really try to learn the material presented to you that day. If you have prepared before class and briefly reviewed the notes afterward, you'll be surprised at how quickly and how well your concentrated study time will progress. You may even find yourself enjoying chemistry!

As you study, continue to fill in the bare spots in your class notes. Write out the definitions of new terms in your notebook. In this way, when it comes time for an exam you should be able to review for it simply from your notes.

At this point you're probably thinking that there isn't enough time to do all the things described above. Actually, the preparation before class and brief review of the notes shortly after class take very little time and will probably save more time than they consume.

Well, you're on your way to an A. There are a few other things that can help you get there. If you possibly can, spend about 30 minutes to an hour at the end of a week to review the week's work. Psychologists have found that a few brief exposures to a subject are more effective at fixing them in the mind than a "cram" session before an exam. The brief time spent at the end of a week can save you hours just before an exam (efficiency!). Try it (you'll like it); it works.

There are some people (you may be one of them) who still have difficulty with chemistry even though they do follow good study habits. Often this is because of weaknesses in their earlier education. If, after following intensive study, you are still fuzzy about something, speak to your teacher about it. Try to clear up these problems before they get worse. Sometimes, by having study sessions with fellow classmates you can help each other over stumbling blocks.

Problem Solving

A stumbling block for many chemistry students is numerical problems. Both the textbook and this study guide have worked-out examples in which the solutions to problems are given in rather great detail. Your instructor will also be showing you how to solve problems. But this is not enough! Learning to solve chemistry problems is like learning to play a musical instrument or drive a car. You only learn by doing. Even if you "understand" how a problem is worked out, you still have to try others yourself to see if you really understand the material sufficiently to solve them. Keep working out problems until you can do them; then you can stop. All the problems in the study guide have answers given. The even-numbered numerical problems in the text also have answers given in Appendix C. Work on these so that you can see whether you are getting them correct.

One of the goals of both the text and this study guide is to teach you how to solve numerical problems. Perhaps this one aspect of chemistry, more than all others, makes you fearful about your fate in the course - you're afraid of the "math." Actually, though, there is very little mathematics involved in solving most of the chemistry problems you will meet. Most of the difficulty comes in trying to interpret a question so that you know what kind of problem you're supposed to solve. In this section we'll go over some basic approaches to solving problems. If you have difficulty with a problem later on, review the ideas presented here and try to apply them. You'll find that they're useful not only in chemistry, but in other areas as well, including problems you encounter in day-to-day living.

"Word problems" always seem to present students with the most difficulty. "What am I supposed to find?" "Where do I begin?" These are the kinds of questions you've probably asked yourself when faced with a word problem. Many people have found the following to be the most effective way of approaching problems of this type.

Step 1. First, preview the problem to get an overview of the question - the "big picture." At this point, don't get bogged down by details. Don't worry about numbers or specific formulas that may be encountered in the question. Read the entire question without trying to analyze it in detail. Remember, at this point you're only interested in getting a view of the whole problem.

Step 2. After you've looked over the entire problem, the next step is to identify what it is you are asked to compute. Look for key phrases such as "Find...." or "How many...." or "What is...." These allow you to know where you're headed in the solution. You might also try to make an educated guess at the magnitude of the answer, although this isn't essential at this point.

Step 3. Now that you know where you're headed, look over the information provided in the question. Don't worry about the numbers yet; simply examine the nature of the preliminary data. Sometimes it's helpful to extract the data from a word problem and tabulate it so that it isn't cluttered with words.

Step 4. Consider next the kinds of calculations that you must perform on the data. Don't worry about the numbers yet. Simply analyze how you must combine the data in the problem to get the answer you want. Be sure you have everything you need. If you use the factor-label method described in Appendix A of the text, you should be able to write simple equality statements such as:

$$1 \text{ ft} = 12 \text{ in.}$$
$$1 \text{ yd} = 3 \text{ ft}$$

Notice that these two statements have the units "ft" in common and provide sufficient information to convert yards to inches, or vice versa. Be sure your equality statements connect all the units so that you have a path from the starting data to the final answer. If a connection between units is missing, you haven't assembled all the relationships that you need to work out the arithmetic. Look for the missing link, either in the data given in the problem or in the knowledge that you're supposed to bring to bear on that kind of problem.

In this step you also must be sure you have any necessary chemical equations or mathematical formulas. Be sure to write them down on paper - don't try to work with them in your head.

Step 5. Well, now you can finally worry about the numbers! At this point all of the necessary information has been compiled and you've decided how you must solve the problem. Now you should go about inserting numbers into formulas or constructing and applying the conversion factors as described in Appendix A of the text. If you've done your preparation in Steps 1 to 4, obtaining an answer in this step should not be difficult.

<u>Step 6</u>.   Take a deep breath, you've done it!   The problem is solved.   As a final point, look at the answer you obtained.   Does it seem reasonable?   Are the units correct?   If so, you're finished.

Time to Begin

Well, it's hoped that the few suggestions presented in this introduction will help you over the hurdles in chemistry.   Move on to the course now, and good luck on getting that A!

1 INTRODUCTION

As its name implies, this chapter is meant to introduce you to the study of chemistry. It begins to lay the foundation for the remainder of the course. If you have had chemistry in high school, perhaps much of the material covered in this chapter will be familiar to you. You can test your knowledge by reviewing the list of new terms at the end of each section in Chapter 1 of the Study Guide and by taking the self-tests below. If you've never taken chemistry before, you should be sure to begin the course properly by gaining a thorough understanding of the topics treated here.

1.1 THE SCIENTIFIC METHOD

Objectives

To understand how science develops through the process of observation, formation of theories, and the design of new experiments that test these theories. You should know the distinction between a law and a theory; between qualitative and quantitative observations.

Review

The scientific method is the procedure that scientists use, either consciously or unconsciously, in their investigation of nature. Data are collected and condensed into laws. Theories are invented in an attempt to explain the laws. The theories suggest new experiments that produce new data, new laws and ultimately new theories. This cycle repeats itself over and over as our understanding of nature grows.

Self-Test (True or False)

1. A law is based on repeated observation.   _____

2. A law is an explanation of a theory.   _____

3. Theories can always be proven to be correct.   _____

4. Laws are often expressed in the form of a mathematical equation.   _____

5. A hypothesis is a tentative law.   _____

6. Numbers are usually associated with qualitative observations.   _____

New Terms

scientific method   law
qualitative observation   hypothesis
quantitative observation   theory
data

1.2   MEASUREMENT

Objectives

To understand that the extent of our knowledge of the world about us is limited by the precision of the measurements that we make. You should be able to recognize the number of significant digits in a number and be able to express the result of a computation to the proper number of significant figures.

Review

Remember that in counting up significant figures in a number, only zeros that are not required for the sole purpose of locating the decimal point should be included. Some examples are:

number	number of significant figures
302	3
0.012	2
2.012	4
0.0120	3

In performing computations with numbers that come from measurements, remember these rules:

1. <u>Multiplication or division</u>. The answer has the same number of significant figures as the least precise factor in the calculation; for example,

$$3.05 \quad\quad x \quad\quad 1.3 \quad\quad = \quad\quad 3.965 = 4.0$$

(3 sig. figures)   (2 sig. figures)   (answer rounded to 2 sig. figures)

2. <u>Addition or subtraction</u>. The number of significant figures in the answer is controlled by the quantity having the largest uncertainty; for example,

this quantity⎫   214.3   (implies uncertainty of ± 0.1)
has largest ⎬→+ 21   (implies uncertainty of ± 1)
uncertainty ⎭   235   (answer has implied uncertainty of ± 1)

Example 1.1

Perform the following arithmetic and express the answer to the proper number of significant figures. All the numbers come from measurement.

$$\frac{(2.500 + 0.10) \times 12.35}{1.468}$$

Solution

We perform the arithmetic within parentheses first.

$$2.500 + 0.10 = 2.60$$

Now our problem is

$$\frac{2.60 \times 12.35}{1.468} = 21.873297$$

Because 2.60 has only three significant figures, the answer is rounded to 21.9.

Exact numbers come from definitions. For instance, 1 mile is exactly 5280 feet, with no uncertainty. Similarly, 1 foot is exactly equal to 12 inches, no more or less. In calculations, these numbers may be assumed to possess any desired number of significant figures.

Example 1.2

A desk was measured to be 34.3 in. along its smallest length. Will it fit through a door that is known to be 2.75 feet wide?

Solution

Let's convert 34.3 in. to feet. We can use the relationship between feet and inches to construct a conversion factor (see Appendix A in the text) that enables us to change the units inches into the units feet.

$$34.3 \text{ in} \left(\frac{1 \text{ ft}}{12 \text{ in}} \right) = 2.86 \text{ ft}$$

Notice that we may assume that both the 1 and the 12 have as many significant figures as we wish. Since there are three significant figures in 34.3, the answer can be expressed to three significant figures. Also note that the desk won't fit through the door!

In Example 1.2 we have cancelled units just as on Page 7 of the text. You should spend some time now to study and review the factor-label method. It is described in detail in Appendix A (p 798) of the text. Although this method may seem foreign to you now, if you learn to apply it, you will find that setting up the arithmetic of chemistry problems is really not very difficult at all.

Self-Test

7. Give the number of significant figures in each of the following.
    (a) 205.3   _____

    (b) 113   _____

(c)   200.0   _____

(d)   0.005   _____

(e)   0.0000700   _____

8.   Evaluate the following expressions to the proper number of significant figures (assume all numbers represent measured quantities).

(a)   2.43 x 1.875 =   _____

(b)   0.017 x 5.968 =   _____

(c)   1.43 x 2.584 x 0.008 =   _____

(d)   12.5 ÷ 2.8 =   _____

(e)   14.34 ÷ 4.780 =   _____

(f)   5.146 + 0.002 =   _____

(g)   5.146 + 0.02 =   _____

(h)   8.08 + 80.8 =   _____

(i)   14.45 + 7.521 + 100.3 =   _____

(j)   2.92 - 8.4 =   _____

9.   Evaluate the following to the proper number of significant figures (assume all numbers represent measured quantities).

(a) $\dfrac{1.43 \times 2.658}{(2.65 + 0.01)}$ _____

(b)   (6.33 x 8.415) + 8.02   _____

New Terms

significant figures   factor-label method
precision   exact numbers
accuracy

1.3   UNITS OF MEASUREMENT

Objectives

To learn the basic SI and metric systems of units and to become familiar with conversions from one unit to another within the metric system. You should also learn to express numbers in scientific notation.

Review

You should familiarize yourself with the SI base units and their symbols. Those that you will encounter in this course are:

Physical quantity	Unit	Symbol
mass	kilogram	kg
length	meter	m
time	second	s
electric current	ampere	A
temperature	kelvin	K
quantity of substance	mole	mol

It is important to learn the eight SI prefixes printed in color in Table 1.2. Remember that when one of these prefixes is used, it modifies the basic unit by the corresponding factor in the first column of the table. For example, nano is a prefix meaning "x 10^{-9}." Therefore, 1 nanogram = 1 x 10^{-9} gram (or, 1 ng = 1 x 10^{-9} g, using the symbols for the prefixes and units). Study Table 1.3 to be sure you understand how to apply the SI prefixes.

Be sure you are well familiar with the units most used for measurements in the laboratory.

Measurement	Unit
length	meter, centimeter, millimeter
mass	gram
volume	liter, milliliter, cubic centimeter

Learn to convert from one unit to another (e.g., liters to milliliters, centimeters to millimeters, etc.). Remember:

$$1 \text{ cm} = 10 \text{ mm}$$
$$1 \text{ liter} = 1000 \text{ ml} = 1000 \text{ cm} \quad (\text{or cc})$$

Most conversions that you will encounter between the English system and the metric system can be handled by remembering the following:

length: 1.00 inch = 2.54 cm
weight: 2.20 lb = 1.00 kg
volume: 1.00 quart = 946 ml

These can provide the cross-over between the two systems of units. For example, to convert 4.0 ft into meters, first change ft to in., then in. to cm, and finally cm to meters.

$$4.0 \text{ ft}\left(\frac{12 \text{ in.}}{1 \text{ ft}}\right)\left(\frac{2.54 \text{ cm}}{1 \text{ in.}}\right)\left(\frac{1 \text{ m}}{100 \text{ cm}}\right) = 1.2 \text{ m}$$

Note the cancellation of units. In the Self-Test, practice using units to guide the arithmetic. Remember, if the units don't cancel properly you will obtain the wrong answer, no matter how much you paid for your calculator!

Review, in Appendix A (Page 799), the writing of numbers in scientific notation. Even though your calculator may handle scientific notation, you should understand the principles of arithmetic operations on numbers expressed in this form. See the following Self-Test for practice.

Self-Test (fill in the blanks)

10. Make the following conversions:

  (a) 150 m = _____ cm

  (b) 27 cm = _____ mm

  (c) 1.50 liter = _____ ml

  (d) 0.002 g = _____ μg

  (e) 100 cm^2 = _____ mm^2 (If necessary, see Example 1.4 on p 10.)

  (f) 253 ml = _____ liter

  (g) 0.143 g = _____ mg

  (h) 1 m^3 = _____ cm^3

  (i) 1 km = _____ cm

  (j) 84 ml = _____ liter

11. Make the following conversions:

  (a) 12.4 in. = _____ cm

  (b) 18.3 cm = _____ in.

  (c) 1.2 mm = _____ in.

  (d) 18.0 m = _____ yd

(e) 1.3 ft^2 = _____ m^2

(f) 1400 cm^3 = _____ qt (assume three significant figures)

(g)   185 km = _____ miles

(h)   2.37 lb = _____ g

(i)   84.0 g = _____ oz

(j)   1.00 ton = _____ kg

12.   Without using a calculator, express the following in scientific notation:

(a)   1,400 = _____ (assume three significant figures)

(b)   275.3 = _____

(c)   0.00307 = _____

(d)   0.00002 = _____

13.   Without using a calculator, write the following in ordinary decimal notation:

(a) 3.0×10^3 = ___3000___

(b) 2.0×10^6 = _____

(c) 1.5×10^{-5} = _____

(d) 34×10^{-7} = _____

(e) 0.025×10^3 = _____

14.   The following are measured quantities. How many significant figures are represented in each?

(a) 3×10^2 cm = _____

(b) 2.00×10^{-3} km = _____

(c) 4.000×10^2 g = _____

15.   Write each of the measurements in Question 14 in ordinary decimal notation.

(a) _____   (b) _____   (c) _____

New Terms

SI	meter
base units	liter
metric system	exponential notation
gram	scientific notation

1.4   MATTER

Objectives

To learn the distinction between weight and mass. You should learn the definition of the term, matter.

Review

Matter has mass and occupies space. The quantity of matter in an object is specified by giving its mass. Mass is the object's resistance to a change in velocity and is a measure of the amount of matter in a particular sample; weight is the force with which an object is attracted to the earth. Mass is measured by comparing the weights of objects on a balance.

New Terms

matter	weight
mass	balance

1.5   PROPERTIES OF MATTER

Objectives

To distinguish between intensive and extensive properties; you should learn the meaning of density and specific gravity, and how to use them in calculations. You should understand the difference between physical properties and chemical properties of matter.

Review

Remember, an intensive property is one that is independent of the size of the sample under consideration. For instance, all samples of pure water, regardless of size, freeze at 32°F; freezing point is an example of an intensive property. Volume is an extensive property - one that depends on sample size. The volumes occupied by different samples of water, for example, are different, depending on the amount of water in each sample.

Melting point and volume are also physical properties; that is, they may be specified without referring to another substance. Chemical properties are always described by relating one substance to another.

Density is a very useful intensive property. It relates mass to volume. It can be used to calculate the volume of a given mass of substance, and vice versa. The three quantities, density (d), mass (m), and volume (V) are related by the equation,

$$d = \frac{m}{V} \qquad (1.1)$$

If you know any two quantities in this equation, you can calculate the third.

Example 1.3

What volume would be occupied by 8.53 g of a substance whose density is 2.54 g/ml?

Solution

You can solve this problem either by Equation 1.1 above or by the cancellation of units. Solving Equation 1.1 for the volume,

$$V = \frac{m}{d}$$

and substituting

$$V = \frac{8.53 \text{ g}}{2.54 \text{ g/ml}}$$

gives

$$V = 3.36 \text{ ml}$$

To solve the problem by unit cancellation you must realize that the density is a conversion factor relating mass and volume.

$$d = \frac{2.54 \text{ g}}{1 \text{ ml}} \qquad \text{or} \qquad \frac{1}{d} = \frac{1 \text{ ml}}{2.54 \text{ g}}$$

Therefore,

$$8.53 \text{ g} \left(\frac{1 \text{ ml}}{2.54 \text{ g}} \right) = 3.36 \text{ ml}$$

Specific gravity is the ratio of the density of a substance to the density of water.

$$\text{sp. gr.} = \frac{d_{substance}}{d_{water}}$$

Remember that specific gravity is unitless. It can be used to calculate density in a variety of units if the density of water in those units is known.

$$d_{substance} = (\text{sp. gr.}) \times d_{water}$$

Self-Test

16. Indicate whether each of the following are intensive (I) or extensive (E) properties:

(a) mass _____   (d) boiling point _____

(b) color _____   (e) density _____

(c) volume _____   (f) specific gravity _____

17. Indicate whether each of the following are physical (P) or chemical (C) properties:

(a) nickel chloride is green. _____

(b) carbon monoxide combines with oxygen to produce carbon dioxide. _____

(c) grain alcohol boils at 78.5°C. _____

(d) ozone (produced in smog) is very reactive toward gasoline vapors. _____

(e) carbohydrates are metabolized in the body to produce carbon dioxide and water. _____

18.   An object with a mass of 14.3 g displaces 5.22 ml of water
       when placed in a graduated cylinder. Calculate the density
       of the object.

       _____

19.   What is the mass of 25.0 ml of an oil if its density is
       0.843 g/ml?

       _____

20.   What volume of alcohol (density = 0.789 g/ml) has a mass of
       18.0 g?

       _____

21.   The density of water is 8.34 lb/gallon, or 62.4 lb/ft^3. The
       specific gravity of sea water is 1.025. What is the density
       of sea water in units of

       (a)   lb/gallon   _____

       (b) lb/ft^3 _____

       (c) g/cm^3 _____

New Terms

extensive properties   specific gravity
intensive properties   physical properties
density   chemical properties

1.6   ELEMENTS, COMPOUNDS, AND MIXTURES

Objectives

       To understand the distinction between these three classes
       of substances.

Review

       Remember, mixtures may be of variable composition, such
as solutions of salt in water. Compounds and elements are always
of fixed (constant) composition. Mixtures may be separated by
physical means into their component compounds. Compounds can
only be separated into elements by chemical reaction. In order of
decreasing complexity we have: mixtures, compounds, elements.
Elements are the simplest substances that are encountered in the
chemistry laboratory. Study Figure 1.9 on Page 18 cf the text.

Self-Test

22. Why do we classify water as a compound?

_____

23. How many elements are presently known? _____

24. Sea water is a mixture. What does this statement tell us?

_____

_____

New Terms

element   thin-layer chromatography
compound   phase
mixture   phase change
physical process   homogeneous
distillation   heterogeneous
chromatography

1.7   THE LAWS OF CONSERVATION OF MASS AND DEFINITE PROPORTIONS

Objectives

To appreciate the historical significance of these two important chemical laws.

Review

These two basic laws of chemistry govern much of our quantitative thinking about chemical reactions.

Self-Test

25. State the law of conservation of mass. _____

_____

26. State the law of definite proportions. _____

_____

27.   Suppose one sample of the compound sodium chloride (table
      salt) contained 23.0 g of the element sodium and 35.5 g of
      the element chlorine. In a different sample, how many
      grams of sodium would be found combined with 71.0 g of
      chlorine?

                                                    _____

New Terms

law of conservation of mass   law of definite composition
law of definite proportions

1.8   THE ATOMIC THEORY OF DALTON

Objectives

To appreciate the historical significance of the development
of the atomic theory. You should understand how Dalton's
theory explains the chemical laws in Section 1.7 and how it
predicts the law of multiple proportions.

Review

Review the postulates of Dalton's theory and the definition
of a molecule. Below is a sample problem dealing with the law of
multiple proportions.

Example 1.4

Two compounds are formed between copper and oxygen.
In one there is 0.290 g of oxygen combined with 2.30 g of copper;
in the other there is 0.466 g of oxygen combined with 1.85 g of
copper. Show that these data demonstrate the law of multiple
proportions.

Solution

You must calculate the weight of one element (let's say
oxygen) combined with the same weight of the other element
(copper) in the two compounds. The data supplied gives you a
chemical equivalence between copper and oxygen in the two compounds.

Compound I   2.30 g copper ~ 0.290 g oxygen
Compound II   1.85 g copper ~ 0.466 g oxygen

The weight of oxygen that would be combined with 1.00 g of copper in each of these compounds is:

Compound I

$$1.00 \text{ g copper} \left(\frac{0.290 \text{ g oxygen}}{2.30 \text{ g copper}} \right) = 0.126 \text{ g oxygen}$$

Compound II

$$1.00 \text{ g copper} \left(\frac{0.466 \text{ g oxygen}}{1.85 \text{ g copper}} \right) = 0.252 \text{ g oxygen}$$

The law of multiple proportions holds if the ratio of these weights of oxygen is a ratio of small whole numbers. Therefore, you set up the ratio,

$$\frac{0.126 \text{ g oxygen}}{0.252 \text{ g oxygen}} = \frac{1}{2}$$

Self-Test

28. In two compounds, each containing 1.00 g of carbon, there are 0.333 g and 0.167 g of hydrogen, respectively. What is the ratio of weights of hydrogen in the two compounds?

29. Phosphorus and oxygen form two compounds. In compound I there are 3.47 g of oxygen combined with 2.68 g of phosphorus; in the other there are 2.82 g of oxygen combined with 3.64 g of phosphorus. What is the ratio of the weights of oxygen that combine with 1.00 g of phosphorus in the two compounds?

30. A 4.00-g sample of cupric bromide was heated, driving off some of the bromine and leaving 2.57 g of cuprous bromide. This cuprous bromide was then decomposed to give bromine and pure copper (1.14 g). Assuming that no copper was lost during these chemical changes, calculate the weights of bromine in the two copper compounds.

(a) Wt. of bromine in cupric bromide   _____

(b) Wt. of bromine in cuprous bromide   _____

(c) What is the ratio of the weights of bromine in the two compounds?   _____

New Terms

molecule   chemical equivalence
law of multiple proportions

1.9   ATOMIC WEIGHTS

Objectives

To understand how a table of atomic weights can be established by comparing the relative weights of the elements
that combine to form compounds of known composition.

Review

The atomic weights that we use are not the weights of
individual atoms, but rather the relative weights of atoms compared to one particular isotope of carbon as a standard.   The
atomic mass of carbon-12 is <u>exactly</u> 12 amu.

New Terms

atomic mass unit   dalton
atomic weight   isotopes

1.10   SYMBOLS, FORMULAS, AND EQUATIONS

Objectives

To begin to become familiar with chemical symbols, formulas, and chemical equations.

Review

Note that in writing the symbol for an element the first
letter is capitalized while the second is not.   The subscripts in a
formula denote the relative numbers of each kind of element contained in that compound.   For example, $Na_2S_4O_7$ contains two
atoms of sodium (Na), four atoms of sulfur (S) and seven atoms
of oxygen (O).   The formula $Ca(NO_3)_2$ shows <u>one</u> calcium atom,
<u>two</u> nitrogen atoms and <u>six</u> oxygen atoms.   The formulas for hy-

drates show the numbers of water molecules trapped in crystals. For example, $CrCl_3 \cdot 6H_2O$ contains six water molecules for each $CrCl_3$.

Chemical equations are used to indicate what occurs during a chemical reaction. An equation is balanced if there are the same number of atoms of each element on the reactant side (left side) of the arrow as there are on the product side (right side). The symbols s = solid, ℓ = liquid, g = gas, and aq = aqueous solution are used sometimes to indicate the phases of reactants and products.

Self-Test

Refer to the alphabetical list of elements on the inside front cover of the textbook to check your answers to Questions 31 and 32.

31.   Write the symbol for each of the following elements.

(a) sodium _____   (f) copper _____

(b) sulfur _____   (g) chlorine _____

(c) oxygen _____   (h) potassium _____

(d) hydrogen _____   (i) magnesium _____

(e) iron _____   (j) carbon _____

32.   What are the names of the following elements?

(a) Sn _____   (f) Cr _____

(b) Br _____   (g) Ag _____

(c) Al _____   (h) As _____

(d) Ca _____   (i) I _____

(e) P _____   (j) He _____

33.   How many atoms of each kind are represented in the formulas below?

(a) KCl _____

(b) NO_2 _____

(c) N_2O_4 _____

(d) $(NH_4)_3PO_4$ _____

(e) $Al_2(SO_4)_3$ _____

(f) $KAl(SO_4)_2 \cdot 12H_2O$ _____

34.   Which of the equations below are <u>not</u> balanced?

(a) $CaO + H_2O \longrightarrow Ca(OH)_2$ _____

(b) $CaCl_2 + H_2SO_4 \longrightarrow HCl + CaSO_4$ _____

(c) $Br_2 + 2NaOH \longrightarrow NaOBr + NaBr + H_2O$ _____

(d) $SO_2 + O_2 \longrightarrow SO_3$ _____

(e) $Al_2(SO_4)_3 + 2CaCl_2 \longrightarrow 2BaSO_4 + AlCl_3$ _____

New Terms

chemical symbol   reactants
chemical formula   products
chemical equation   coefficients
hydrate   balanced equation

1.11   ENERGY

Objectives

To learn the difference between kinetic energy and potential energy. You should understand the difference between heat and temperature and the units used to express them. You should also learn the concept of specific heat.

Review

Kinetic energy is associated with motion ($KE = \frac{1}{2}mv^2$); potential energy is associated with the distance of separation between particles that either attract or repel one another. Remember that if there is neither an attraction nor repulsion between two objects, there are no potential energy changes when the objects move toward or away from each other.

Temperature is an intensive quantity that determines the direction of heat flow (hot $\longrightarrow$ cold). The Celsius scale (or centigrade scale) defines 0°C as the melting point of ice (which is the same as the freezing point of water). The boiling point of water is defined as 100°C. The kelvin scale has -273.15°C as its zero point. Remember that the size of the degree unit is the same on the Celsius and Kelvin scales. A temperature change of one Celsius degree is the same as a change of one kelvin degree. You should be able to convert from one of these temperature scales to another.

Various units used to express energy are the erg, joule (J), kilojoule (kJ), calorie (cal), and kilocalorie (kcal). Remember these conversions:

$$1 \text{ cal } = 4.1840 \text{ J}$$
$$1 \text{ kcal } = 4.1840 \text{ kJ}$$

Review the definition of specific heat and Example 1.11 on Page 29 of the text.

Self-Test

35. Perform the following conversions:

(a)   25°C = _____ K

(b)   -30°C = _____ K

(c)   350 K = _____ °C

(d)   77 K = _____ °C

(e)   4.0°C = _____ °F

(f)   50°F = _____ °C

(g)   254 kJ = _____ kcal

(h)   32.5 kcal = _____ kJ

36. A 1.50-g piece of gold absorbed 0.162 cal when its temperature was raised by 3.50°C. Calculate the specific heat of gold in the units.

(a) cal/g °C   _____

(b) J/g °C   _____

37. How is the joule defined in terms of the SI base units?

_____

38. How is the temperature of the surroundings affected when a system undergoes an exothermic change? _____

_____

New Terms

energy   temperature
kinetic energy   Fahrenheit scale
potential energy   Celsius scale
exothermic   Kelvin scale
endothermic   calorie
law of conservation of energy   kilocalorie
erg   specific heat
joule

Answers to Self-Test Questions

1. true   2. false   3. false   4. true   5. false   6. false   7. (a) 4
(b) 3   (c) 4   (d) 1   (e) 3   8.(a) 4.56   (b) 0.10   (c) 0.03
(d) 4.5   (e) 3.000   (f) 5.148   (g) 5.17   (h) 88.9   (i) 122.3
(j) −5.5   9.(a) 1.43   (b) 61.3   10.(a) 15,000 cm   (b) 270 mm
(c) 1500 ml   (d) 2000 μg   (e) 10,000 mm^2 (f) 0.253 liter
(g) 143 mg   (h) 1,000,000 cm^3 (i) 100,000 cm   (j) 0.084 liter
11.(a) 31.5 cm   (b) 7.20 in.   (c) 0.047 in.   (d) 19.7 yd
(e) 0.12 m^3 (f) 1.48 qt   (g) 115 mi   (h) 1080 g (to 3 sig. fig.)
(i)   2.96 oz   (j) 909 kg   12.(a) 1.40×10^3 (b) 2.753×10^2
(c) 3.07×10^{-3} (d) 2×10^{-5} 13.(b) 2,000,000   (c) 0.000015
(d) 0.0000034   (e) 25   14.(a) 1   (b) 3   (c) 4   15.(a) 300 cm
(b) 0.00200 km   (c) 400.0 g   16.(a) E   (b) I   (c) E   (d) I
(e) I   (f) I   17.(a) P   (b) C   (c) P   (d) C   (e) C
18. 2.74 g/ml   19. 21.1 g   20. 22.8 ml   21.(a) 8.55 lb/gallon
(b) 64.0 lb/ft^3 (c) 1.025 g/cm^3 22. All samples contain hydrogen and oxygen in the same proportions.   23. 106   24. It can be
of variable composition and it can be separated into components
by physical processes.   25. Mass is neither created nor destroyed
during a chemical reaction.   26. In a compound the elements are
always present in the same proportions by mass.   27. 46.0 g of
sodium   28. 2 to 1   29. (g of O in I)/(g of O in II) = 1.29/0.775
= 1.66 = 5/3   30.(a) 2.86 g   (b) 1.43 g   (c) 2 to 1   31.(a) Na
(b) S   (c) O   (d) H   (e) Fe   (f) Cu   (g) Cl   (h) K   (i) Mg
(j) C   32.(a) tin   (b) bromine   (c) aluminum   (d) calcium
(e) phosphorus   (f) chromium   (g) silver   (h) arsenic   (i) iodine
(j) helium   33.(a) K, 1; Cl, 1   (b) N, 1; O, 2   (c) N, 2; O, 4
(d) N, 3; H, 12; P, 1; O, 4   (e) Al, 2; S, 3; O, 12   (f) K, 1;

Al, 1; S, 2; O, 20; H, 24   34.(a) balanced   (b) unbalanced
(c) balanced   (d) unbalanced   (e) unbalanced   35.(a) 298 K
(b) 243 K   (c) 77°C   (d) -196°C   (e) 39.2°F   (f) 10°C
(g) 60.7 kcal   (h) 136 kJ   36.(a) 0.0309 cal/g °C
(b) 0.129 J/g °C   37. 1 J = 1 kg m^2/s^2 38. The temperature
rises.

2 STOICHIOMETRY: CHEMICAL ARITHMETIC

This chapter deals with calculations involving quantities of chemical substances, either combined together in a compound or reacting with one another in a chemical reaction. It is very important that you thoroughly understand the concepts developed here because they are necessary in future discussions in other chapters. This is particularly true of the mole concept discussed in Section 2.1. Most students who develop difficulties with chemistry have not really acquired a genuine "feel" for the mole. Therefore, this is really a very important chapter for you to master well. A little extra time spent here may save you a lot of grief later.

2.1 THE MOLE

Objectives

To learn to think of the mole as the "chemist's dozen." You should develop the ability to translate between ratios of numbers of atoms and molecules that combine and ratios of numbers of moles of these substances that combine. You should also be able to convert from moles to grams, and vice versa. You should learn to use Avogadro's number where appropriate.

Review

Like the dozen (12) or the gross (144), the mole represents a fixed number of objects.   These can be atoms, molecules, or <u>anything</u> we wish to consider.   One mole of an element or compound contains a large enough quantity of atoms or molecules to be worked with in a laboratory.   The most important feature of the mole concept, however, is that in a chemical reaction (for instance, the formation of a compound from its elements) the RATIO in which atoms combine is <u>exactly</u> the same as the RATIO in which moles of atoms combine.   When $CrCl_3$ is formed, three Cl atoms are required for each Cr atom.   In laboratory-sized quantities, three moles of Cl atoms are required for each one mole of Cr atoms.

A concept that many students find difficult to grasp is that we obtain one mole of $CrCl_3$ from three moles of Cl and one mole of Cr.

$$1 \text{ mol Cr} + 3 \text{ mol Cl} \longrightarrow 1 \text{ mol of } CrCl_3$$

If this troubles you, consider this analogy:

$$1 \text{ doz frames} + 3 \text{ doz wheels} \longrightarrow 1 \text{ doz tricycles}$$

Both equations involve precisely the same kind of reasoning.

You should learn to convert between laboratory units of grams and chemical units of moles.   Remember that one mole of any element has a mass in grams numerically equal to the element's atomic weight.   For example, the atomic weight of fluorine is 19.0; 1 mol of F = 19.0 g F.

Avogadro's number (6.022×10^{23} things = 1 mol things) is used <u>only</u> when you must translate sizes (mass, volume, length, etc.) between the large world that we work in (with balances, graduated cylinders, and rulers) to the submicroscopic world of individual atoms and molecules.

<u>Self-Test</u>

1. How many moles of F must react with one mole of S to form:

    (a) SF_2 _____   (b) SF_4 _____   (c) SF_6 _____

2. What mole ratio of carbon to hydrogen is found in propane, C_3H_8?

    _____

3. How many moles of Cl must react with 0.50 mol of P to form PCl_5?

    _____

4. How many moles of oxygen atoms are there in 1.20 mol of Fe_3O_4?

    _____

5. How many moles of iron (Fe) atoms are there in 1.20 mol of Fe_3O_4?

    _____

6. Oxygen occurs as molecules of O_2. How many moles of O atoms are there in 1.40 mol of O_2?

    _____

7. How many moles of O_2 would be needed to prepare 5.0 mol of N_2O_4?

    _____

8. How many moles of Cl_2 would be needed to prepare 3.0 mol of PCl_5?

    _____

9. What is the weight in grams (to three significant figures) of one mole of:

    (a) carbon atoms   _____

    (b) potassium atoms   _____

    (c) calcium atoms   _____

    (d) nickel atoms   _____

    (e) bromine atoms   _____

10. How many moles of atoms are there in:

    (a) 32.1 g S   _____

    (b) 46.0 g Na   _____

    (c) 12.5 g Ag   _____

    (d) 3.50 g N   _____

11. How many grams does each of the following weigh?

    (a) 1.00 mol Mn   _____

    (b) 0.455 mol Al   _____

    (c) 1.34 mol Ba   _____

    (d) 2.14 mol Zn   _____

12. How many grams of each element are present in 0.250 mol of Al_2O_3?

    _____

13. How many grams of O are needed to react with 0.300 mol S to form SO_3?

_____

14. How many grams of Cl must react with 10.0 g C to form C_2Cl_4?

_____

15. How many <u>atoms</u> are there in:

(a) 1.00 mol S   _____

(b) 1.00 mol O_2 _____

(c) 0.341 mol P   _____

(d) 1.85 mol Cl_2 _____

16. What is the weight in grams of 1 atom of Si?_____

17. What is the weight in grams of 1 molecule of O_2?

_____

18. How many atoms are there in 5.27 g of Na? _____

19. How many grams do 1.40×10^{21} atoms of silver weigh?

_____

<u>New Terms</u>

mole   Avogadro's number
mol

2.2   MOLECULAR WEIGHTS AND FORMULA WEIGHTS

<u>Objectives</u>

To learn to calculate molecular weights and formula weights.

<u>Review</u>

The molecular weight, or formula weight, is simply the sum of all of the atomic weights of all of the atoms in the formula of a compound. For example, the molecular weight of glucose, $C_6H_{12}O_6$, is:

$$
\begin{array}{lll}
\text{carbon} & 6 \times 12.01 = & 72.06 \\
\text{hydrogen} & 12 \times 1.008 = & 12.10 \\
\text{oxygen} & \underline{6 \times 16.00 =} & \underline{96.00} \\
& \text{formula weight of } C_6H_{12}O_6 = & 180.16
\end{array}
$$

Remember that 1 mol of a compound = 6.022×10^{23} formula units and has a weight in grams numerically equal to the formula weight. Thus, 1 mol $C_6H_{12}O_6$ = 180.16 g.

The term formula weight is always used for compounds that are ionic. (At this time, however, you are not expected to know which compounds are ionic.)

Self-Test

20. Calculate the formula weights of:

    (a) KNO_3 (potassium nitrate)   _____

    (b) $CO(NH_2)_2$ (urea)   _____

    (c) $C_9H_8O_4$ (aspirin)   _____

    (d) CCl_4 (carbon tetrachloride)   _____

    (e) NaOCl (bleach)   _____

21. How many moles are there in each of the following?

    (a) 14.3 g of $NaC_{18}H_{35}O_2$ (soap)   _____

    (b) 142 g of $(CH_3)_2CO$ (acetone - used in nail polish remover)

                            _____

22. What is the weight in grams of:

    (a) 1.00 mol of $CaCl_2$ (calcium chloride)   _____

    (b) 0.0250 mol of $CHCl_3$ (chloroform)   _____

    (c) 3.46 mol of Fe_2O_3 (rust)   _____

    (d) 14.2 mol of $KAl(SO_4)_2 \cdot 12H_2O$ (alum)   _____

New Terms

molecular weight   electron
formula weight   ion
formula unit   ionic compound

2.3   PERCENTAGE COMPOSITION

Objectives

To calculate the percentage composition of a compound from its formula. You should be able to calculate the weight of a given element in a compound given its formula.

Review

The weight percent of an element in a compound is calculated by dividing the weight of that element in the compound by the molecular weight and then multiplying by 100. The percent Cl in CCl_4 is:

$$\% \text{ Cl in } CCl_4 = \left(\frac{\text{wt Cl in } CCl_4}{\text{M.W. } CCl_4} \right) \times 100$$

$$= \left(\frac{(4)(35.5)}{12.0 + (4)(35.5)} \right) \times 100$$

$$= \left(\frac{142}{154} \right) \times 100$$

$$= 92.2\%$$

To calculate the weight of a given element in a sample of a compound you multiply the weight of the sample by the fraction of the compound that is the desired element. For example, to calculate the weight of chlorine in 85.0 g of CCl_4 we multiply the weight of the sample (85.0 g) by the fraction of CCl_4 that is chlorine.

$$\text{fraction that is Cl} = \frac{\text{wt Cl in } CCl_4}{\text{M.W. } CCl_4}$$

$$= \frac{142 \text{ g Cl}}{154 \text{ g } CCl_4}$$

The weight of Cl in the sample, then, is

$$\text{wt Cl in sample} = 85.0 \text{ g } \cancel{CCl_4} \left(\frac{142 \text{ g Cl}}{154 \text{ g } \cancel{CCl_4}} \right) = 78.4 \text{ g Cl}$$

Self-Test

23. Calculate the percentage compositions of each of the following:
    (a) $NaNO_3$ _____

    (b) $Ca(HCO_3)_2$ _____

    (c) $NiCl_2$ _____

24. Calculate the weight of Ag in 22.0 g of AgCl. _____

25. Calculate the weight of sulfate, SO_4, in 12.3 g of $BaSO_4$.

    _____

26. In a chemical analysis a 3.14-g sample known to contain $CuSO_4$ and $CuCl_2$ was dissolved in water and treated with $Ba(NO_3)_2$. Solid $BaSO_4$ was formed, which was filtered from the solution, dried and weighed. The $BaSO_4$ weighed 2.58 g.

    (a) What weight of SO_4 was in the $BaSO_4$? _____

    (b) What was the weight percent SO_4 in the original sample?

    _____

New Terms

percentage composition

2.4   CHEMICAL FORMULAS

Objectives

To learn what types of information different kinds of chemical formulas provide.

Review

The <u>simplest formula</u> only gives the relative numbers of atoms in the compound. The <u>molecular formula</u> also gives the actual number of each element in a molecule of the compound. The <u>structural formula</u> describes the way in which the atoms in a molecule are linked together.

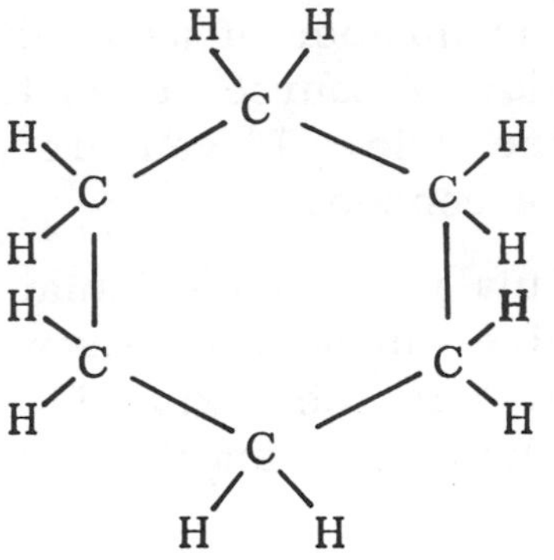

structural formula
for cyclohexane

C_6H_{12}

molecular formula
for cyclohexane

CH_2

empirical formula
for cyclohexane

Self-Test

27. What is the molecular formula and the empirical formula for
the compound whose structural formula is

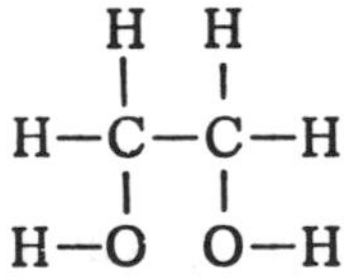

(ethylene glycol - antifreeze)

_____

New Terms

simplest formula
empirical formula

molecular formula
structural formula

2.5   EMPIRICAL FORMULAS

Objectives

To calculate empirical formulas from percentage composition,
or from the weights of elements combined together in a
compound.

Review

An empirical formula gives the atom ratio of elements in the compound; it also gives the ratio of the number of moles of each element. The atom ratio and mole ratio, of course, have to be identical, based on the definition of the mole. If you are unsure about this, you should review the mole concept.

To determine an empirical formula you must calculate the number of moles of each element combined in a given sample of the compound, as shown in Example 2.12 in the text. The ratio of moles gives the atom ratio. The following is another example:

Example 2.1

A 4.00-g sample of a copper-bromine compound was decomposed, yielding 1.14 g of pure copper. What is the empirical formula of the compound?

Solution

First, we must have the weight of Br combined with the Cu. This can be obtained in this case as the difference between the total weight of compound and the weight of copper in the compound.

$$4.00 \text{ g} - 1.14 \text{ g} = 2.86 \text{ g Br}$$

Thus, there were 2.86 g of Br combined with 1.14 g of Cu in the original sample.

Next, we calculate the number of moles of Cu and Br.

$$1.14 \text{ g Cu} \left(\frac{1 \text{ mol Cu}}{63.5 \text{ g Cu}} \right) = 0.0179 \text{ mol Cu}$$

$$2.86 \text{ g Br} \left(\frac{1 \text{ mol Br}}{79.9 \text{ g Br}} \right) = 0.0358 \text{ mol Br}$$

Finally, set up the formula as $Cu_{0.0179}Br_{0.0358}$ and divide through by the smallest subscript to obtain whole numbers.

$$Cu_{\frac{0.0179}{0.0179}} Br_{\frac{0.0358}{0.0179}} = CuBr_2$$

(The data for this question came from Question 30 in Chapter 1 of the Study Guide.)

When an analysis is presented in the form of percentage composition by weight, you can write the percentages of each element as weights by assuming 100 g of compound in the sample. Thus 100 g of a compound that is 50% sulfur and 50% oxygen would contain 50 g S and 50 g O.

Self-Test

28. Determine the empirical formulas for the following compounds.

(a) 7.86 g potassium, 7.14 g chlorine   _____

(b) 37.9 g Na, 17.1 g P   _____

(c) 31.9% K, 29.0% Cl, 39.2% O   _____

(d) 69.6% Mn, 30.4% O   _____

29. A 1.500-g sample containing only nickel and bromine was dissolved in water and allowed to react with silver nitrate. The reaction gave 2.578 g of AgBr.

(a) How many grams of Br were in the AgBr?   _____

(b) How many grams of Br were in the nickel-bromine compound?   _____

(c) How many grams of Ni were in the nickel-bromine compound?   _____

(d) What is the empirical formula of the nickel-bromine compound?   _____

New Terms

2.6   MOLECULAR FORMULAS

Objectives

To determine the molecular formula of a substance, given its empirical formula and its molecular weight.

Review

The molecular formula is always a multiple of the empirical formula. For example, benzene has a molecular formula of C_6H_6; its empirical formula is CH. The empirical formula occurs "six times" in the molecular formula; that is, the subscripts in the empirical formula of benzene must each be multiplied by six to give the molecular formula. The molecular weight of C_6H_6 is 78; the formula weight of CH is 13. The number of times the empirical formula is repeated is obtained by dividing the molecular weight by the empirical formula weight.

$$\frac{78}{13} = 6$$

Self-Test

30. Determine molecular formulas for the following compounds.

empirical formula	molecular weight	molecular formula
CH_2O	180	_____
CH_3	30	_____
P_2O_3	220	_____
HgCl	472.2	_____

New Terms

2.7 BALANCING CHEMICAL EQUATIONS

Objectives

To write balanced chemical equations.

Review

You should keep in mind the advice that writing a balanced equation is a two-step process. First, write the unbalanced equation with correct formulas for each of the reactants and products. Then, balance the equation. At this point in the course you must balance the equation by inspection; that is, you juggle the coefficients (the numbers preceding the formulas) to make the number of atoms of each kind the same on both sides of the equation. Remember, once you have written correct formulas for the reactants and products you do not change the subscripts in the formulas. Also remember that a properly balanced equation has the smallest whole number set of coefficients. Practice balancing the equations in the Self-Test below.

Self-Test

31. Balance the following equations.

(a) $CuSO_4 + Al \longrightarrow Al_2(SO_4)_3 + Cu$

_____

(b) $KClO_3 \longrightarrow KCl + O_2$ _____

(c) $CO(NH_2)_2 + H_2O \longrightarrow CO_2 + NH_3$

_____

(d) $PCl_5 + H_2O \longrightarrow H_3PO_4 + HCl$

_____

(e) $N_2O_5 + H_2O \longrightarrow HNO_3$

_____

New Terms

2.8   CALCULATIONS BASED ON CHEMICAL EQUATIONS

Objectives

To use a balanced chemical equation to perform calculations involving quantities of substances entering into chemical reaction.

Review

A chemical equation such as

$$2H_2 + O_2 \longrightarrow 2H_2O$$

tells us about reacting molecules. It gives information about what happens on a submicroscopic, atomic scale. The mole concept allows us to expand this information up to laboratory-sized quantities. Whatever ratio exists between atoms or molecules of reactants and products, the same ratio exists between moles of reactants and products. For example, the equation above tells us that for every two molecules of H_2 that react, one molecule of O_2 will also react. Scaling this to laboratory-sized quantities, we can say that for every two moles of H_2 that react, one mole of O_2 will react.

In dealing with chemical equations, chemists do their thinking in terms of moles. Let's use the equation above as an example. If a chemist knew that he had 0.40 mol of O_2, he would look at the equation and realize immediately that he would require 0.80 mol of H_2. The equation tells him that twice as many moles of H_2 must react as O_2. Similarly, he would also conclude that he would be able to obtain 0.80 mol of H_2O. The equation tells him that however moles of H_2 are consumed, the same number of moles of H_2O will be formed.

The purpose of doing these calculations is to be able to measure proper quantities of reactants (or products) in the laboratory. However, we cannot measure moles directly; instead we can only measure mass (i.e., grams). We therefore must translate back and forth between laboratory units (grams) and chemical units (moles).

There are a number of worked-out examples presented in the text illustrating the types of calculations you might encounter. Let's look at another.

Example 2.2

How many grams of H_2 are needed to react completely with 4.75 g of Fe_2O_3 according to the equation,

$$Fe_2O_3 + 3H_2 \longrightarrow 2Fe + 3H_2O$$

Solution

All reasoning involving H_2 and Fe_2O_3 must take place in chemical units. The solution of this problem can be diagrammed as shown below.

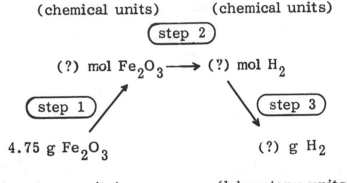

Steps 1 and 3 involve translation between grams and moles (this was covered in Sections 2.1 and 2.2).

Step 2 requires the use of the coefficients in the balanced equation.

Step 1 - Translation

$$4.75 \text{ g Fe}_2\text{O}_3 \times \left(\frac{1 \text{ mol Fe}_2\text{O}_3}{159.6 \text{ g Fe}_2\text{O}_3} \right) = 0.0298 \text{ mol Fe}_2\text{O}_3$$

Step 2 - The coefficients in the equation allow us to establish the chemical equivalency,

$$1 \text{ mol Fe}_2\text{O}_3 \sim 3 \text{ mol H}_2$$

This is then used to construct a conversion factor so that we can calculate the number of moles of H_2 required.

$$0.0298 \text{ mol Fe}_2\text{O}_3 \times \left(\frac{3 \text{ mol H}_2}{1 \text{ mol Fe}_2\text{O}_3} \right) = 0.0894 \text{ mol H}_2$$

Step 3 - Translation

$$0.0894 \text{ mol H}_2 \times \left(\frac{2.02 \text{ g H}_2}{1 \text{ mol H}_2} \right) = 0.180 \text{ g H}_2$$

Self-Test

32. The reaction between hydrazine, N_2H_4, and hydrogen peroxide, H_2O_2, has been used to power rockets.

$$N_2H_4 + 2H_2O_2 \longrightarrow N_2 + 4H_2O$$

(a) How many moles of N_2H_4 are required to react with 8.00 mol of H_2O_2?

_____

(b) How many moles of N_2 will be formed from 8.00 mol of H_2O_2?

_____

(c) How many moles of water will be formed from 8.00 mol of H_2O_2?

_____

(d) How many grams of water will be formed when 3.00 mol of N_2H_4 react?

_____

(e) How many moles of N_2 will be formed when 500 g of H_2O_2 react?

_____

(f) How many grams of H_2O_2 are required to react with 1000 g of N_2H_4?

_____

New Terms

2.9 LIMITING-REACTANT CALCULATIONS

Objectives

To calculate the amount of products formed when arbitrary amounts of reactants are mixed.

Review

These calculations deal with chemical reactions in which substances are simply mixed together without prior regard for maintaining the proper mole ratios between reactants. In these cases, all reactants usually are not consumed completely; one or more of them remains in excess. The amount of product formed in these situations is controlled by the reactant that is completely used up (the limiting reactant), since once it is gone no more product is able to form. In this type of problem, you first determine the limiting reactant and then base your calculation of the amount of product formed on the amount of the limiting reactant available.

Example 2.3

Zinc and oxygen combine to produce zinc oxide according to the equation,

$$2Zn + O_2 \longrightarrow 2ZnO$$

How much ZnO will be formed if 14.3 g of Zn are mixed with 3.72 g of O_2?

Solution

First, we calculate how many moles of Zn and O_2 are in the mixture.

$$14.3 \text{ g Zn} \times \left(\frac{1 \text{ mol Zn}}{65.4 \text{ g Zn}} \right) = 0.219 \text{ mol Zn}$$

$$3.72 \text{ g O}_2 \times \left(\frac{1 \text{ mol O}_2}{32.0 \text{ g O}_2} \right) = 0.116 \text{ mol O}_2$$

Next, we determine the limiting reactant by choosing one reactant and calculating the amount of the other required to give complete reaction. It doesn't matter which we choose, so let's work with the Zn.

$$0.290 \text{ mol Zn} \times \left(\frac{1 \text{ mol O}_2}{2 \text{ mol Zn}} \right) \sim 0.109 \text{ mol O}_2 \text{ required}$$
to react with all the Zn

Notice that we have more O_2 than we need. This means that some O_2 will be left over and all of the Zn will react; zinc is the limiting reactant.

Once the limiting reactant is established we use it to calculate the amount of product that will be formed.

$$0.219 \text{ mol Zn} \times \left(\frac{1 \text{ mol ZnO}}{1 \text{ mol Zn}} \right) \times \left(\frac{81.4 \text{ g ZnO}}{1 \text{ mol ZnO}} \right) \sim 17.8 \text{ g ZnO}$$

The weight of ZnO formed is 17.8 g.

Self-Test

33. Based on the equation,
$$N_2H_4 + 2H_2O_2 \longrightarrow N_2 + 4H_2O$$

   (a) How many moles of N_2 will be formed if 600 g of N_2H_4 are mixed with 1200 g of H_2O_2? _____

   (b) How many grams of H_2O will be produced if 83.5 g of N_2H_4 are mixed with 175 g of H_2O_2? _____

New Terms

limiting reactant

2.10 THEORETICAL YIELD AND PERCENTAGE YIELD

Objectives

To see that not all reactions produce the theoretical maximum amount of product. You should learn the definitions of theoretical yield and percentage yield.

Review

The theoretical yield is the amount of product that would be produced if the reactants were to combine to the maximum extent possible. We calculate the theoretical yield from the limiting reactant following the procedure in the last section. In simpler cases we calculate it as the maximum amount of product formed from a particular reactant as in Section 2.8. The percentage yield is calculated as shown on Page 52 of the text.

Self-Test

34. Glucose, $C_6H_{12}O_6$, is converted to ethyl alcohol, C_2H_5OH, and CO_2 by fermentation,
$$C_6H_{12}O_6 \longrightarrow 2C_2H_5OH + 2CO_2$$
Starting with 200 g of glucose,

(a)   What is the theoretical yield of ethyl alcohol? _____

(b)   If 97.3 g of C_2H_5OH was obtained, what was the percentage yield?

_____

New Terms

actual yield   percentage yield
theoretical yield

2.11   MOLAR CONCENTRATION

Objectives

To learn the meanings of some terms used in discussing solutions. You should learn the meaning of molar concentration and how to use it as a conversion factor relating amounts of solute and volumes of solutions.

Review

In general, in a solution the solvent is present in largest amount and the solute (or solutes) are present in a lesser amount. The proportions of solute and solvent are specified by giving the solution's concentration. Molar concentration, or molarity, is a convenient concentration unit for dispensing measured amounts of solute dissolved in a solution. Remember the units of molarity.

$$\text{molarity} = \frac{\text{moles of solute}}{\text{liter of solution}}$$

To calculate molarity we need to know the number of moles of the solute and the volume of the solution in which it is dissolved.

If you know the molarity of a solution you can use it to calculate the number of moles of solute in a specified volume of the solution, or to calculate the volume of solution needed to contain a specified number of moles of solute. Think of molarity as a conversion factor. You should be able to translate a label such as 2.50 M H_2SO_4 into these factors.

$$\frac{2.50 \text{ mol } H_2SO_4}{1 \text{ liter solution}} \qquad\qquad \frac{1 \text{ liter solution}}{2.50 \text{ mol } H_2SO_4}$$

Examples 2.25 and 2.26 illustrate how to use this kind of conversion factor.

In preparing a solution of a known molarity, remember that the solvent (water, for example) is added to the solute until the desired final volume of solution is reached. To prepare 500 ml of solution, for example, we do <u>not</u> just add 500 ml of water to the solute. First the solute is dissolved in a small amount of water, and then more water is added until the final volume is 500 ml.

Self-Test

35. What is the molarity of the following solutions?

(a) 0.350 mol $NaHCO_3$ in 0.400 liter of solution _____

(b) 0.250 mol KCl in 200 ml of solution _____

(c) 15.6 g of $MgCl_2$ in 300 ml of solution _____

(d) 1.85 g of $AgNO_3$ in 75.0 ml of solution _____

36. How many moles of $CaCl_2$ are in

(a) 1.15 liter of 0.840 M $CaCl_2$ solution? _____

(b) 325 ml of 0.150 M $CaCl_2$ solution? _____

37. What volume (in milliliters) of 3.00 M NH_3 solution contains

(a) 1.35 mol of NH_3? _____

(b) 21.4 g of NH_3? _____

38. How many grams of KNO_3 are needed to prepare 750 ml of 0.200 M KNO_3 solution?

_____

New Terms

solute

solvent

concentration

molar concentration

molarity

molar

Answers to Self-Test Questions

1.(a) 2   (b) 4   (c) 6   2. 3 mol C to 8 mol H   3. 2.5 mol Cl
4. 4.80 mol O   5. 3.60 mol Fe   6. 2.80 mol O   7. 10 mol O_2
8. 7.5 mol Cl_2 9.(a) 12.0 g   (b) 39.1 g   (c) 40.1 g   (d) 58.7 g
(e) 79.9 g   10.(a) 1.00 mol S   (b) 2.00 mol Na   (c) 0.116 mol Ag
(d) 0.250 mol N   11.(a) 54.9 g Mn   (b) 12.3 g Al   (c) 184 g Ba
(d) 140 g Zn   12. 13.5 g Al, 12.0 g O   13. 14.4 g O   14. 59.1 g
Cl   15.(a) 6.02×10^{23} (b) 1.20×10^{24} (c) 2.05×10^{23}
(d) 2.23×10^{24} 16. 4.66×10^{-23} g   17. 5.31×10^{-23} g
18. 1.38×10^{23} atoms   19. 0.251 g   20.(a) 101.1   (b) 60.0
(c) 180   (d) 154   (e) 74.5   21.(a) 0.0467   (b) 2.45   22.(a) 111 g
(b) 2.98 g   (c) 553 g   (d) 6.74×10^3 g   23.(a) 27.1% Na, 16.5%
N, 56.5% O   (b) 24.7% Ca, 1.24% H, 14.8% C, 59.2% O   (c) 45.3%
Ni, 54.7% Cl   24. 16.6 g Ag   25. 5.06 g SO_4 26.(a) 1.06 g SO_4
(b) (1.06 g/3.14 g) x 100 = 33.8% SO_4 27. molecular formula =
$C_2H_6O_2$; empirical formula = CH_3O 28.(a) KCl   (b) Na_3P
(c) $KClO_3$ (d) Mn_2O_3 29.(a) 1.097 g Br   (b) 1.097 g Br
(c) 0.403 g Ni   (d) $NiCl_2$ 30. $C_6H_{12}O_6$, C_2H_6, P_4O_6, Hg_2Cl_2
31.(a) $3CuSO_4 + 2Al \longrightarrow Al_2(SO_4)_3 + 3Cu$
 (b) $2KClO_3 \longrightarrow 2KCl + 3 O_2$
 (c) $CO(NH_2)_2 + H_2O \longrightarrow CO_2 + 2NH_3$
 (d) $PCl_5 + 4H_2O \longrightarrow H_3PO_4 + 5HCl$
 (e) $N_2O_5 + H_2O \longrightarrow 2HNO_3$
32.(a) 4.00 mol N_2H_4 (b) 4.00 mol N_2H_4 (c) 16.0 mol H_2O
 (d) 216 g H_2O (e) 7.35 mol H_2 (f) 2.12×10^3 g H_2O_2
33.(a) 17.6 mol N_2 (H_2O_2 is limiting reactant)
 (b) 188 g H_2O (H_2O_2 is limiting reactant)
34.(a) 102 g   (b) 95.4%
35.(a) 0.875 M   (b) 1.25 M   (c) 0.546 M   (d) 0.145 M
36.(a) 0.966 mol   (b) 0.0488 mol
37.(a) 450 ml   (b) 419 ml
38. 15.2 g KNO_3

3 ATOMIC STRUCTURE AND THE PERIODIC TABLE

Atoms are not indivisible particles as Dalton had originally envisioned them. Instead, they are composed of simpler particles, neutrons and positively charged protons in the nucleus of the atom and negatively charged electrons surrounding the nucleus. The first half of this chapter follows the historical developments that have led to the presently accepted theory about atomic structure. The second part of the chapter describes the electronic structure of the atom and how certain observed properties can be related to atomic structure. This is very important because the chemical behavior of atoms is controlled by their electronic structures.

3.1 THE ELECTRICAL NATURE OF MATTER

Objectives

To understand how experimental evidence was accumulated that showed matter to be electrical in nature.

Review

Faraday's experiments on electrolysis showed that chemical reactions could be caused by electricity.

J. J. Thomson's experiments using the cathode ray tube showed electrons to be fundamental particles. He measured the charge-to-mass ratio for the electron.

The first self-test will be found after Section 3.7. Read and study the first seven sections of the chapter, using the Study Guide in the usual way, before attempting to answer the questions in the self-test.

New Terms

electron
gas discharge tube
electrode
cathode
anode

cathode rays
charge-to-mass ratio
fundamental particle
coulomb

3.2   THE CHARGE ON THE ELECTRON

Objectives

To understand how the charge on the electron was measured.

Review

R. A. Millikan determined the charge on the electron by measuring the charge on oil drops that had picked up electrons. The charge on the oil drops was always a multiple of -1.60×10^{-19} coulombs, and Millikan reasoned that this value must be equal to the electron's charge. (Your instructor probably doesn't expect you to memorize this number - to be sure, though, you should ask.)

New Terms

3.3   POSITIVE PARTICLES AND THE MASS SPECTROMETER

Objectives

To understand that atoms must also contain positive particles and that these positive particles are much heavier
than the electron. You should learn that charges on particles are expressed in multiples of the charge on the
electron.

Review

The mass spectrometer is a device used to measure the
charge-to-mass ratio of positive particles (positive ions). These
ions always have much smaller e/m ratios than the electron, which
tells us that they are much heavier than the electron. The
largest e/m ratio is observed for the hydrogen ion, which is simply a proton. The proton is a fundamental particle.

All atoms of elements other than hydrogen contain more
than one proton, and all the atoms of a given element have the
same number of protons. This number is called the element's
atomic number.

The electron is assigned a charge of -1; the proton has a
charge of +1. This is because charge is gained or lost by atoms
when they gain or lose electrons. A relative charge of -1 really
corresponds to an actual charge of -1.60×10^{-19} coulombs. You
will almost always deal with relative charges.

New Terms

ion   proton
mass spectrometer   atomic number

3.4   RADIOACTIVITY

Objectives

To learn the types of radioactivity shown by certain kinds
of atoms and understand that this phenomenon also shows

that there are particles simpler than the atom.

Review

Three basic types of radioactivity are observed:

α-rays composed of He^{2+} ions (α-particles)
β-rays composed of electrons (β-particles)
γ-rays composed of very penetrating radiation similar to
    X rays

Since these emissions come spontaneously from atoms of certain substances, these atoms must be composed of smaller, simpler particles.

New Terms

alpha particle   gamma rays
beta particle   radioactivity

3.5   THE NUCLEAR ATOM

Objectives

To see how experiments led to the idea that the atom has a tiny, very dense positive nucleus.

Review

E. Rutherford concluded that the atom must possess a very tiny nucleus containing all of the positive charge in the atom and nearly all its mass. This is the only way he could account for the scattering of α-particles at large angles from thin metal foils, and the fact that most of the α-particles passed through the foil nearly unaffected.

New Terms

nucleus

3.6   THE NEUTRON

Objectives

   To examine the properties of the fundamental particle
called the neutron.

Review

   Neutrons are particles of zero charge and of mass almost
the same as the proton. You should review the properties of the
proton, neutron and electron in Table 3.1.

New Terms

3.7   ISOTOPES

Objectives

   To learn the meaning of the term isotope. You should
learn how to write the symbol for a given isotope of an
element and how to calculate the average atomic mass from
the actual isotopic masses and their relative abundances.

Review

   Isotopes of the same element have the same atomic number
(number of protons) but different numbers of neutrons. Remember that when writing the symbol for an isotope, the atomic number, Z, is a left subscript and the mass number, A (the sum of
protons plus neutrons), is a left superscript. The number of
neutrons is A - Z. For example,

$$^{70}_{32}\text{Ge}$$
Z = 32 (32 protons)
A = 70
A - Z = 38 (38 neutrons)

   The relative atomic masses discussed previously are actually average atomic masses. Example 3.1 in the text and the ex-

ample below illustrate how the average atomic mass can be calculated from fractional abundances and accurate relative isotopic masses. (Remember, the actual mass of an isotope is not the same as its mass number.)

Example 3.1

Naturally occurring chlorine is composed of a mixture of 75.53% ^{35}Cl, and 24.47% ^{37}Cl. These have isotopic masses of 34.969 and 36.966 amu, respectively. Calculate the average atomic mass of chlorine.

Solution

Multiply the mass of each isotope by its fractional abundance (obtained from percent by dividing by 100). Then add the results to get the average atomic mass.

$$^{35}Cl \qquad (34.969 \text{ amu})(0.7553) = 26.41$$
$$^{37}Cl \qquad (36.966 \text{ amu})(0.2447) = \underline{9.05}$$
$$\text{Total} = 35.46$$

New Terms

mass number
radioactive decay

Self-Test on Sections 3.1 to 3.7

Try to answer these questions without referring back to the text or to the review material in this book.

1. (True or False)

    (a) Faraday's experiments permitted the determination of the charge-to-mass ratio of the electron.

    (b) The cathode ray tube used by Thomson is similar to a television picture tube.

    (c) Cathode rays have different properties for different samples of matter.

(d) The charge on the electron was measured by experiments using charged oil droplets.   _____

(e) From the data obtained from Thomson's cathode ray tube experiments and Millikan's experiments, the mass of the electron could be calculated.   _____

(f) The charge-to-mass ratio for positive particles is always larger than the charge-to-mass ratio for the electron.   _____

(g) An alpha particle is the same as an electron.   _____

(h) Gamma rays are not particles, but instead are high energy light waves.   _____

(i) In the mass spectrometer the positive particle with the largest e/m ratio is the proton.   _____

(j) The diameter of the nucleus of an atom is approximately 1/100,000 of the diameter of the atom.   _____

(k) The number of protons in the nucleus of an atom is given by the mass number.   _____

(l) Isotopes of a given element have the same number of protons but differ in the number of neutrons.   _____

2. How many protons are there in these atoms?

(a) $^{32}_{16}$S _____   (b) $^{192}_{77}$Ir _____   (c) $^{39}_{19}$K _____

3. How many neutrons are there in these atoms?

(a) $^{108}_{47}$Ag _____   (b) $^{209}_{83}$Bi _____   (c) $^{19}_{9}$F _____

4. Give the number of protons, neutrons, and electrons in the following.

Ion	No. of protons	No. of neutrons	No. of electrons
$^{126}_{53}$I			
$^{118}_{50}$Sn^{2+}			
$^{79}_{34}$Se^{2-}			

5. (Fill in the blanks)

   (a) An atom that has acquired a charge by the gain or loss of electrons is called _____

   (b) The relative charge on an α-particle is _____

   (c) The relative charge on a γ-ray is _____

   (d) When a current of 1 ampere flows for 1 second the amount of charge that passes a given point in a wire is called _____

6. Antimony occurs in nature as a mixture of two isotopes, 57.25% ^{121}Sb with a mass of 120.904 amu and 42.75% ^{123}Sb with a mass of 122.904 amu. What is the average atomic mass of Sb? _____

3.8 THE PERIODIC LAW AND THE PERIODIC TABLE

Objectives

To understand the rationale behind the structure of the periodic table and the locations of elements within it. You should learn the nomenclature that applies to the periodic table and the names applied to different sets of elements.

Review

The periodic table is without a doubt one of the most useful devices available to a chemist or a chemistry student. Trends in a large number of chemical and physical properties can be correlated with the positions of the elements within this table. Your ability to use the periodic table effectively depends on how well you understand its construction.

Remember that the elements are arranged in horizontal rows (called periods) in order of increasing atomic number. The periodic law states that when arranged in this manner the elements exhibit a periodic recurrence of properties. An important feature of the periodic table is that elements with similar chemical properties are arranged in vertical columns (called groups).

This section introduces you to a set of nomenclature associated with the periodic table. The important terms are given in boldface type in the text and are listed on the following page. You should learn the meanings of these terms. After you've studied them, take the following self-test.

7. Which of the following are representative elements: Cl, Fe, Cu, Na, Xe?

   _____

8. Which of these are transition elements: As, Hg, Ti, Ge, Sr?

   _____

9. Which of these is a halogen: S, Sn, Br, Na, Mg?

   _____

10. Which of these is a noble gas: O, Ne, Na, Ca, Zn?

    _____

11. Which of these is an alkali metal: Li, B, C, F, Xe?

    _____

12. Which of these is an alkaline earth metal: Zn, C, Cs, Ba, Kr?

    _____

13. Which of these are inner transition elements: Np, Ru, F, As, Pm?

    _____

14. Which of these is a metalloid: S, Ni, Ge, He, Mg?

    _____

15. Which are metals: Sr, Si, Cr, Ce, U, P?

    _____

16. Which are nonmetals: S, Ga, P, Pr, I, K?

    _____

17. What is the more common term that means "family of elements"?

    _____

18. What kind of elements are malleable and ductile?

    _____

New Terms

periodic law

group

period

representative elements

transition elements

inner transition elements

lanthanides

actinides

rare earths

alkali metals

alkaline earth metals

halogens

noble gases

metal

metalloid

nonmetals

ductility

malleability

3.9   ELECTROMAGNETIC RADIATION AND ATOMIC SPECTRA

Objectives

To learn some interrelated properties of light waves and to understand that the light emitted by an atom that has been "energized" is composed of only a relatively few colors, rather than an entire rainbow.

Review

Light (electromagnetic radiation) travels as waves through a vacuum at a constant speed (c) equal to 3.00×10^8 m/s. An even more precise value of c can be found on the inside rear cover of the text. The intensity of the wave is its amplitude. The product of the wave's frequency (ν) and its wavelength (λ) is equal to c.

$$\lambda \cdot \nu = c$$

The SI unit of frequency is the hertz: $1 \text{ Hz} = 1 \text{ s}^{-1}$

A white-hot object like the sun or an incandescent lamp emits light of all colors to give a continuous spectrum. Excited atoms emit only certain wavelengths (colors) and produce a line spectrum. It is possible to find an equation that allows the calculation of the wavelengths of the lines. An important aspect of this equation is that it involves the difference between the reciprocals of squares of integers. The occurrence of these integers provides the clue to the electronic structure of the atom.

Self-Test

19.   A typical radar transmitter emits microwaves with a frequency of 9300 MHz (megahertz). What is the wavelength of these waves expressed in meters?

_____

20.   Sodium emits yellow light at a wavelength of 589 nm. What is its frequency?

_____

21.   The frequency of a green light wave is 5.49×10^{14} Hz. What is its wavelength in

   (a)   centimeters   _____

   (b)   nanometers   _____

22.   What is the wavelength, in nanometers, of the second line of the Lyman series in the atomic spectrum of hydrogen?

_____

23.   How can an atom be excited so that it emits its characteristic atomic spectrum? _____

24.   In the Rydberg equation, if $n_1 = 5$, what values can n_2 have?

_____

25.   Each element has its own characteristic x-ray spectrum that can be related to the element's atomic number (True or False?)

_____

New Terms

electronic structure   hertz
electromagnetic radiation   continuous spectrum
amplitude   atomic emission spectrum
wavelength   line spectrum
frequency   Rydberg equation

3.10   THE BOHR THEORY OF THE HYDROGEN ATOM

Objectives

   To show that the introduction of the idea of quantized energy levels in the atom permitted the explanation of atomic spectra. You should learn the relationship between frequency of light and energy.

Review

Planck had shown that the energy in a beam of light is proportional to the frequency of the light wave. Remember that $E = h\nu$.

The significance of Bohr's theory was that it introduced for the first time the idea that in an atom the electron is only permitted to have certain energies; intermediate energies are forbidden. Electrons change energy by going from one energy level to another. Energy is absorbed by an atom when one of its electrons is raised from one energy level to a higher one, and energy is released when the electron falls from one energy level to a lower one. Energy levels can be identified by the value of a quantum number.

This section also illustrates how complex theories are tested. From the postulates of the theory an equation is derived, in this case an equation that can be used to calculate the wavelengths of lines in the atomic spectrum. This theoretical equation is compared to an equation based solely on the experimental data. If the equations match, it is taken as evidence for the validity of the theory; if they don't, the theory must be wrong. Bohr's success with hydrogen indicated he was on the right trail. The failure of his theory to predict the wavelengths of spectral lines of atoms more complicated than hydrogen demonstrated that there was a basic flaw somewhere in the theory.

Self-Test

26. What is the energy, in joules, of a photon having

(a) a frequency of 4.50×10^{15} Hz?   _____

(b) a wavelength of 589 nm?   _____

27. (Multiple choice). The existence of line spectra demonstrates that
(a) only certain electrons in atoms can be excited
(b) the electrons in an atom can have only certain specific energies
(c) Planck's equation doesn't always hold true
(d) white light is composed of many wavelengths

(e)   none of these are correct   _____

28.   (Multiple choice).   Bohr's theory

(a)   proved that the electron travels in circular orbits about the nucleus
(b)   concluded that the radius of an orbit was inversely proportional to the quantum number, n
(c)   was successful in explaining the Rydberg equation for hydrogen
(d)   states that an electron gains energy when it moves from an orbit with a given value of n to an orbit with a smaller value of n
(e)   none of the above apply to Bohr's theory   _____

New Terms

photon   energy level
quanta   quantum number
Planck's constant

3.11   WAVE MECHANICS

Objectives

To understand that matter, like light waves, has wavelike properties. You should understand the phenomenon of diffraction and the meanings of the terms, wave function and orbital. You should also learn the names and permissible values of the quantum numbers used to identify electron orbitals.

Review

As predicted by de Broglie, it has been shown experimentally that matter has wave properties. Proof lies in the diffraction of particles. Diffraction results from constructive and destructive interference of waves.

Wave mechanics is the name of the theory that treats the electron as a wave. Solution of a wave equation gives a set of wave functions, ψ. Each wave function describes an atomic orbital that has a characteristic energy and that corresponds to a region around the nucleus where we are likely to find the electron. An orbital is identified by a set of three quantum numbers: n, ℓ, and m. Review the values permitted for these quantum numbers on Page 85 of the text.

$$n = 1, 2, \ldots,$$

$$\ell = 0, 1, 2, \ldots, n-1$$

$$m = 0, \pm 1, \pm 2, \ldots, \pm \ell$$

Remember that subshells are identified by their value of ℓ.

ℓ	0	1	2	3
letter designation	s	p	d	f

Before moving on, examine the energy level diagram in Figure 3.21. Note that within a shell the energy of the subshells vary as: s < p < d < f. Also note that an s subshell consists of one orbital; a p subshell, three orbitals; a d subshell, five orbitals; and an f subshell, seven.

Self-Test

29. What are the values of n, ℓ, and m for each orbital in a 2p subshell?

_____

30. What values of m are allowed in the following?

   (a) 2s subshell   _____

   (b) 3d subshell   _____

   (c) 5f subshell   _____

   (d) 4p subshell   _____

31. Give the proper subshell designation corresponding to the following sets of quantum numbers.

   (a) n = 3, ℓ = 1   _____

   (b) n = 4, ℓ = 3   _____

   (c) n = 4, ℓ = 2   _____

   (d) n = 2, ℓ = 0   _____

32. Why don't we see wave properties for large particles like cars and baseballs?

_____

New Terms

wave mechanics
quantum mechanics
diffraction
diffraction pattern
standing wave
node
wave function

orbital
shell
principal quantum number
azimuthal quantum number
magnetic quantum number
ground state

3.12   ELECTRON SPIN AND THE PAULI EXCLUSION PRINCIPLE

Objectives

To see that the electron behaves as if it were spinning about its axis like a top. You should learn that the Pauli exclusion principle limits the number of electrons per orbital to two. You should learn how electron spin influences the magnetic properties of substances.

Review

The electron behaves like a tiny electromagnet, implying that it is spinning about its axis. There are two values of the spin quantum number s, $+\frac{1}{2}$ and $-\frac{1}{2}$, corresponding to two directions of rotation.

The Pauli exclusion principle requires that any two electrons in an atom have different sets of values for its four quantum numbers: n, ℓ, m and s. If the first three are identical for two electrons, the electrons must spin in opposite directions.

The maximum number of electrons that can be placed in a given orbital is two, and they must have opposite spins. The maximum electron population per subshell is:

subshell	max population
s	2
p	6
d	10
f	14

It sometimes helps to remember this if you realize that the numbers 2, 6, 10, 14 form an arithmetic progression, each successive number being four larger than the one before it.

You should review the magnetic properties of substances as they are determined by the electron's spin.

Self-Test

33. (Multiple choice). One electron in an atom has the quantum numbers: $n = 3$, $\ell = 2$, $m = -1$, $s = \frac{1}{2}$. Which of the following is <u>not</u> a possible set of quantum numbers for a second electron in this same atom?

(a) $n = 1$, $\ell = 0$, $m = 0$, $s = -\frac{1}{2}$
(b) $n = 2$, $\ell = 1$, $m = -1$, $s = \frac{1}{2}$
(c) $n = 3$, $\ell = 2$, $m = -1$, $s = \frac{1}{2}$
(d) $n = 3$, $\ell = 1$, $m = -1$, $s = \frac{1}{2}$
(e) $n = 3$, $\ell = 2$, $m = 0$, $s = \frac{1}{2}$ _____

34. (Fill in the blanks). The maximum number of electrons in

(a) the 2s subshell is   _____

(b) the 3p subshell is   _____

(c) the 6g subshell is   _____

35. A neutral potassium atom must be paramagnetic. Why?

_____

New Terms

spin quantum number   paramagnetic
Pauli exclusion principle   ferromagnetic
diamagnetic

3.13   THE ELECTRON CONFIGURATIONS OF THE ELEMENTS

Objectives

To write electron configurations for the elements. You should learn both the conventional notation (e.g., $1s^2$...) as well as how to construct an orbital diagram.

Review

The number of electrons in a given subshell is specified by writing the subshell designation with the number of electrons indicated as an exponent. Thus $3p^4$ indicates four electrons in a 3p subshell.

Subshells in an atom become populated starting with the lowest energy level first. The sequence in which subshells become filled is determined by the energy level diagram in Figure 3.21 (Page 87 of the text).

When writing orbital diagrams, arrows are used to indicate electrons (head up for one direction of spin, and head down for the other). For example, the orbital diagram for boron (Z = 5) is

$$\text{B} \quad \underset{1s}{\uparrow\downarrow} \quad \underset{2s}{\uparrow\downarrow} \quad \underset{\underline{\quad\quad 2p \quad\quad}}{\uparrow \quad \underline{\quad} \quad \underline{\quad}}$$

or
$$\text{B} \quad [\text{He}] \quad \underset{2s}{\uparrow\downarrow} \quad \underset{\underline{\quad\quad 2p \quad\quad}}{\uparrow \quad \underline{\quad} \quad \underline{\quad}}$$

Notice that all the orbitals of the 2p subshell are shown, even though only one of them is populated by an electron. Remember that [He] stands for the filled noble gas core. In a similar fashion we would write the orbital diagram for Ca as

$$\text{Ca} \quad [\text{Ar}] \quad \underset{4s}{\uparrow\downarrow}$$

When more than one electron occupies a p, d or f subshell, Hund's rule applies, which tells us that for an atom in its ground state the electrons are spread out over the orbitals as much as possible with their spins in the same direction. For example, the orbital diagram for phosphorus is

$$\text{P} \quad [\text{Ne}] \quad \underset{3s}{\uparrow\downarrow} \quad \underset{\underline{\quad\quad 3p \quad\quad}}{\uparrow \quad \uparrow \quad \uparrow}$$

A phenomenon that has some important consequences in terms of chemical properties is that half-filled and filled subshells are extra stable. This causes Cr and Cu to have unexpected electron configurations.

New Terms

electronic structure   paired electrons
electron configuration   core electrons
orbital diagram   Hund's rule

3.14   THE PERIODIC TABLE AND ELECTRON CONFIGURATIONS

Objectives

To learn to use the periodic table to deduce the electron configuration of an element.

Review

The structure of the periodic table is a direct consequence of the order in which subshells are filled and, as described in the text, you can use the periodic table to help you write electron configurations.

Example 3.2

Write the electron configuration of germanium (Z = 32).

Solution

There are 32 electrons in the atom. Moving from left to right across successive periods we fill:

$1s^2$ (period 1)
$2s^2$ (period 2 - Groups IA and IIA)
$2p^6$ (period 2 - Groups IIIA through the noble gases)
$3s^2$ (period 3 - Groups IA and IIA)
$3p^6$ (period 3 - Groups IIIA through the noble gases)
$4s^2$ (period 4 - Group IA and IIA)
$3d^{10}$ (period 4 - first row of transition elements)
$4p^2$ (period 4 - Groups IIIA and IVA)

Writing all this together:

Ge $1s^2 2s^2 2p^6 3s^2 3p^6 4s^2 3d^{10} 4p^2$

Some people prefer to write all subshells of a given shell together.

$$\text{Ge} \quad 1s^2 2s^2 2p^6 3s^2 3p^6 3d^{10} 4s^2 4p^2$$

Showing the noble gas core and those electrons outside it, we can also write

$$\text{Ge} \quad [\text{Ar}] \quad 3d^{10} 4s^2 4p^2$$

If we only wished to know the configuration of the outer shell, we could write

$$\text{Ge} \quad 4s^2 4p^2$$

This can be obtained without having to write the entire configuration, as described in Example 3.11.

Self-Test (Sections 3.13 and 3.14)

36. Give the complete electron configuration of

    (a) Mg _____

    (b) Cl _____

    (c) Tc _____

    (d) Ni _____

37. Give the abbreviated electron configuration (showing the noble gas core) of

    (a) Si _____

    (b) V _____

38. Give the electron configuration of the outer shell of

    (a) Arsenic _____

    (b) Iodine _____

39. Construct orbital diagrams for the following:

    (a) Si   (c) Fe

    (b) Ca   (d) S

New Terms

3.15   THE SPATIAL DISTRIBUTION OF ELECTRONS

Objectives

     To understand how wave mechanics describes the spatial distribution of electrons (i.e., where the electrons are likely to be found).

Review

     Because of the Heisenberg uncertainty principle, wave mechanics describes the probability of finding the electron at points around the nucleus. The electron is viewed as being smeared out throughout the volume of the atom. Regions where the probability of finding the electron is high are said to have a high electron density.

     The shapes of the probability distributions for different types of orbitals will be important in discussion of bonding in Chapter 5. Remember that s orbitals are spherical; p orbitals are dumbbell shaped. You should also know that the three orbitals in a p subshell are oriented at 90° to each other.

     Notice that 2s and higher s orbitals contain nodes (where the electron density drops to zero), as do 3p and higher p orbitals. The important point, however, is that s orbitals have an overall spherical shape while the p orbitals tend to "point" in specific directions. Also notice that as the value of n increases, the size of the orbital increases.

New Terms

uncertainty principle	electron cloud
probability distribution	charge cloud
electron density	

3.16   THE VARIATION OF PROPERTIES
WITH ATOMIC STRUCTURE

Objectives

To be able to predict trends in atomic and ionic size, ionization energy, and electron affinity.

Review

Atomic and ionic sizes have traditionally been specified in units of angstroms (Å): $1 \text{ Å} = 10^{-8} \text{ cm} = 10^{-10} \text{ m}$. Learn the conversions at the bottom of Page 97 which relate the angstrom to the preferred SI units nonometers and picometers.

Atomic size increases going down a group in the periodic table because the outer shell orbitals become larger and the effective nuclear charge that they experience remains nearly the same. Atomic size decreases from left to right in a period because the effective nuclear charge felt by the outer electrons increases.

Remember that the lanthanide contraction causes atoms of the elements immediately following lanthanum to be nearly the same size as those above them. This makes these elements very dense.

Ionization energy is the energy needed to remove an electron from an isolated atom or ion.

Electron affinity is the energy released (or absorbed) when an electron is added to a gaseous atom. Students often find this term confusing. Perhaps a better name for it would have been "electron attachment energy." Unfortunately, however, scientists choose to call this energy the electron affinity, so you have to become accustomed to the name.

As mentioned earlier, within the periodic table, atomic size generally decreases from left to right in a period and increases from top to bottom within a group. Variations in ionization energy and electron affinity generally parallel variations in size. Large atoms have low ionization energies and low electron affinities while small atoms have high ionization energies and high electron affinities. The variation of properties within the periodic table is summarized on the next page.

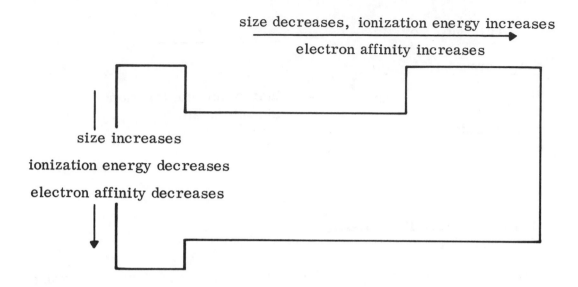

size decreases, ionization energy increases

electron affinity increases

size increases

ionization energy decreases

electron affinity decreases

Remember that the second ionization energy is always larger than the first, the third larger than the second, and so forth. Also remember that the second electron affinity is always an endothermic quantity.

Remember that positive ions are always smaller than the neutral atoms from which they are formed and that an atom gets larger when it becomes a negative ion.

Self-Test

40. The atomic radius of a bromine atom is 1.14 Å. What is this radius expressed in

    (a) nanometers _____ (b) picometers _____ ?

41. Which is the largest atom: C, N, Si, P? _____

42. Which atom has the largest ionization energy: B, C, Al, Si?

    _____

43. Which atom has the smallest electron affinity: B, C, Al, Si?

    _____

44. In each pair, choose the one with the larger radius.

    (a) Na or Na$^+$ _____

    (b) Cl or Cl$^-$ _____

(c) Mn^{2+} or Mn^{3+} _____

New Terms

angstrom
atomic radius
ionic radius
effective nuclear charge

lanthanide contraction
ionization energy
electron affinity

Answers to Self-Test Questions

1.(a) False (b) True (c) False (d) True (e) True (f) False
(g) False (h) True (i) True (j) True (k) False (l) True
2.(a) 16 (b) 77 (c) 19 3.(a) 61 (b) 126 (c) 10 4. 53 protons, 73 neutrons, 53 electrons; 50 protons, 68 neutrons, 48
electrons; 34 protons, 45 neutrons, 36 electrons 5.(a) ion
(b) 2+ (c) zero (d) coulomb 6. 121.8 amu 7. Cl, Na, Xe
8. Hg, Ti 9. Br 10. Ne 11. Li 12. Ba 13. Np, Pm 14. Ge
15. Sr, Cr, Ce, U 16. S, P, I 17. Group 18. metals
19. 3.2×10^{-2} m 20. 5.09×10^{14} Hz 21.(a) 5.46×10^{-5} cm
(b) 546 nm 22. 103 nm 23. electric discharge or heating it in a
flame 24. $n_2 = 6, 7, 8, \ldots, \infty$ 25. True 26.(a) 2.98×10^{-18} J
(b) 3.37×10^{-19} J 27. b 28. c 29. $n = 2$, $\ell = 1$, $m = -1$;
$n = 2$, $\ell = 1$, $m = 0$; $n = 2$, $\ell = 1$, $m = 1$ 30.(a) 0 (b) 0, ± 1,
± 2 (c) 0, ± 1, ± 2, ± 3 (d) 0, ± 1 31.(a) 3p (b) 4f (c) 4d
(d) 2s 32. Their large masses give them very small wavelengths.
33. c 34.(a) 2 (b) 6 (c) 18 35. It has an odd number of
electrons, so they all cannot be paired. 36.(a) Mg $1s^2 2s^2 2p^6 3s^2$
(b) Cl $1s^2 2s^2 2p^6 3s^2 3p^5$ (c) Tc $1s^2 2s^2 2p^6 3s^2 3p^6 3d^{10} 4s^2 4p^6 4d^5 5s^2$
(d) Ni $1s^2 2s^2 2p^6 3s^2 3p^6 3d^8 4s^2$ 37.(a) Si [Ne] $3s^2 3p^2$
(b) V [Ar] $3d^3 4s^2$ 38.(a) As $4s^2 4p^3$ (b) I $5s^2 5p^5$
39.(a) Si [Ne] $\underline{\uparrow\downarrow}$ $\underline{\uparrow}$ $\underline{\uparrow}$ $\underline{}$ (b) Ca [Ar] $\underline{\uparrow\downarrow}$
 3s 3p 4s

(c) Fe [Ar] $\underline{\uparrow\downarrow}$ $\underline{\uparrow\downarrow}$ $\underline{\uparrow}$ $\underline{\uparrow}$ $\underline{\uparrow}$ $\underline{\uparrow}$
 4s 3d

(d) S [Ne] $\underline{\uparrow\downarrow}$ $\underline{\uparrow\downarrow}$ $\underline{\uparrow}$ $\underline{\uparrow}$ 40.(a) 0.114 nm (b) 114 pm
 3s 3p
41. Si 42. C 43. Al 44.(a) Na (b) Cl^- (d) Mn^{2+}

4 CHEMICAL BONDING: GENERAL CONCEPTS

Chemical bonds are what hold atoms together in compounds. The discussion of chemical bonding in your text is divided between two chapters. This first chapter examines some general concepts and provides a simplified treatment of the subject. Here you should learn the types of chemical bonding and how they relate to the electronic structures of the atoms that are bonded. In Chapter 5 you will see how some of the more modern bonding theories account for the formation of bonds and the shapes of molecules.

4.1 LEWIS SYMBOLS

Objectives

      To learn how to write the Lewis symbol, or dot symbol, for atoms of the representative elements.

Review

      Remember that it is the outer shell (valence shell) electrons that are involved in the formation of chemical bonds. Lewis symbols are simply a bookkeeping device that is used to keep track of the valence electrons during bond formation. Dots (or some other symbol such as an x or a circle) are used to represent valence electrons. Review Table 4.1 in the text before working on the following Self-Test.

Self-Test

1. Write Lewis symbols for:

    (a) Si _____   (c) Sr _____

    (b) Cl _____   (d) As _____

New Terms

valence shell   Lewis structure
Lewis symbol   electron-dot formula

4.2   THE IONIC BOND

Objectives

        To understand what an ionic bond is, how it is formed,
    and the types of elements that form ionic bonds.

Review

        Remember that the ionic bond is formed by electron trans-
    fer. This produces ions that attract each other because of their
    opposite charges. Among the representative elements atoms tend
    to gain or lose electrons until they achieve an electron configura-
    tion that is identical to a noble gas (see Table 4.2). A noble
    gas atom has eight electrons in its outer shell, hence the octet
    rule which states that atoms tend to gain or lose electrons until
    there are eight electrons in the outer shell. You should be able
    to diagram the formation of an ionic compound using Lewis sym-
    bols. The Self-Test at the end of this section provides some
    practice.

        Among the transition elements the octet rule doesn't always
    hold - other relatively stable electron configurations also exist.
    Review Table 4.3 in your text.

        Ionic compounds also exist that contain polyatomic ions.
    Learn the formulas, charges, and names of the ions in Table 4.4.
    When writing the formulas and compounds containing these ions
    remember that the ratio of positive to negative ions must be
    chosen to give a neutral formula. For instance, the compound

containing Fe^{3+} and CO_3^{2-} has the formula, $Fe_2(CO_3)_3$. A simple way to get this formula is to use the number of charges on one ion as the subscript on the other.

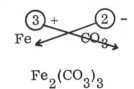

gives $Fe_2(CO_3)_3$

This system works for writing the formula for any ionic compound. Just remember to reduce the subscripts to the smallest set of whole numbers. For example,

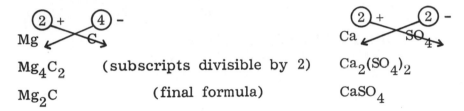

Mg_4C_2 (subscripts divisible by 2) $Ca_2(SO_4)_2$

Mg_2C (final formula) $CaSO_4$

Self-Test

2. Use Lewis symbols to diagram the formation of an ionic bond between:

   (a) K and Cl _____

   (b) Na and O _____

3. Write the pseudonoble gas configuration for Zn^{2+}.

   _____

4. Write the formulas for the ionic compounds formed from the ions:

   (a) Cr^{2+} and PO_4^{3-} _____

   (b) Cu^{2+} and SO_4^{2-} _____

   (c) Ga^{3+} and ClO_2^{-} _____

   (d) Cr^{3+} and $C_2O_4^{2-}$ _____

   (e) NH_4^{+} and S^{2-} _____

5. What are the names of the polyatomic ions in Question 4?

   (a) _____

(b) _____

(c) _____

(d) _____

(e) _____

New Terms

ionic bond   octet rule
electrovalent bond   post-transition element
cation   pseudonoble gas configuration
anion   polyatomic ion

4.3   FACTORS INFLUENCING THE FORMATION OF IONIC COMPOUNDS

Objectives

To examine the factors that favor ionic bonding and to learn what gives rise to the stability of ionic compounds.

Review

An ionic compound is stable because of the very large lattice energy that is released when the ions come together to produce the ionic solid. If it were not for this lattice energy, ionic compounds would not exist. Thus, in the gas phase the ions, $Li^+(g)$ and $F^-(g)$, are less stable (i.e., of higher energy) than $Li(g)$ and $F(g)$ and therefore $Li(g)$ and $F(g)$ would not react spontaneously to produce the ions.

Ionic compounds tend to form most readily when metals of low ionization energy and electron affinity react with nonmetals of high ionization energy and electron affinity, and when the lattice energy of the resultant compound is high. Therefore, the metals on the extreme left of the periodic table (Groups IA and IIA) and the nonmetals on the extreme upper right tend to form ionic compounds.

Self-Test

6. What basic principle permits us to analyze the formation of an ionic compound using a Born-Haber cycle?

   _____

7. Why isn't LiF composed of Li^{2+} and F^{2-} ions instead of Li^+ and F^- ions?

   _____

   _____

   _____

8. The term stability is generally associated with low _____ .

New Terms

lattice energy
Born-Haber cycle

4.4   THE COVALENT BOND

Objectives

To learn how the covalent bond is formed. You should learn how to draw Lewis structures for covalent molecules. This should include molecules and polyatomic ions that contain single, double and triple bonds.

Review

Covalent bonding is favored when the difference between the ionization energies and electron affinities of the combining atoms is not large. Sharing a pair of electrons between two atoms lowers their energy, as shown in Figure 4.4. Learn the meanings of the terms bond length and bond energy. A single covalent bond consists of one pair of electrons shared between two atoms. Double and triple bonds consist of two and three shared pairs. Pairs of electrons in bonds are usually represented by a dash.

By electron sharing, atoms normally complete their octets. An exception is hydrogen, which never has more than two electrons in its valence shell. Compounds of beryllium and boron are also exceptions. The rules we use to draw Lewis structures often do not apply to them.

Study the rules for drawing Lewis structures on Page 119, and study Examples 4.5 - 4.8.

<u>Self-Test</u>

9. Using the Lewis symbols for the elements Si, P, S and Cl predict the formulas for the simplest compounds these elements would form with hydrogen and write their Lewis structures.

10. Write electron-dot formulas for the following (the central atom is written first in each formula).

(a) $SiCl_4$ (d) SO_4^{2-}

(b) ClF_3 (e) CO

(c) SCl_2 (f) ClO_2^-

<u>New Terms</u>

covalent bond   single bond
bond energy   double bond
bond length   triple bond

4.5   BOND ORDER AND SOME BOND PROPERTIES

<u>Objectives</u>

To learn how three bond properties, bond length, bond energy, and vibrational frequency, are related to the electron density between two atoms.

Review

Bond order is the number of electron pairs shared between two atoms. Remember that as the bond order increases,

  (a) bond length decreases
  (b) bond energy, which in a sense is the "bond strength," increases
  (c) vibration frequency increases

Self-Test

11. The bond length in carbon monoxide, CO, is 1.13 Å. Judging from the data in Table 4.4 of your text, what can you say about the bond order in CO?

_____

12. Consider these compounds:

$$
\begin{array}{ccc}
\underset{\text{(I)}}{\overset{\displaystyle H \quad H}{\underset{\displaystyle H}{\overset{|\quad\;|}{\underset{|}{H-C-N-H}}}}}
&
\underset{\text{(II)}}{\overset{\displaystyle H}{\underset{\displaystyle H}{\overset{|}{\underset{|}{H-C-C\equiv N:}}}}}
&
\underset{\text{(III)}}{\overset{\displaystyle H \quad H}{\overset{|\quad\;|}{H-C=N-N-H}}}
\end{array}
$$

  (a) Which would have the longest carbon-nitrogen bond?

                                              _____

  (b) Which would have the largest carbon-nitrogen vibrational frequency?

                                              _____

  (c) Which would have the largest carbon-nitrogen bond energy?

                                              _____

  (d) What would be the approximate carbon-nitrogen bond length in compound (III)?

                                              _____

13. What kind of information is obtained from the infrared absorption spectrum of a molecule?

                                              _____

New Terms

bond order
vibrational frequency

4.6   RESONANCE

Objectives

To learn the meaning of the term resonance and to learn
to draw resonance structures where applicable.

Review

Remember that the actual structure of a resonance hybrid
never corresponds to any of the resonance forms that you draw
for the molecule or ion; it is always something in between. Resonance structures arise when there is more than one reasonable
way of distributing electron pairs in a molecule. The NO_3^- ion,
for example, is a resonance hybrid of three contributing structures:

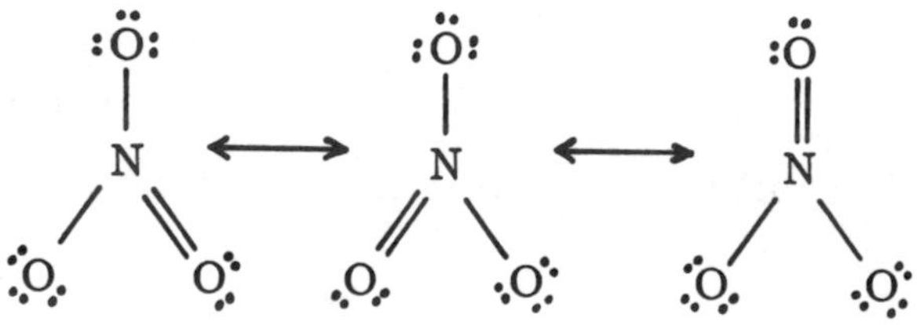

When you draw the dot structure for a molecule or ion and
find that you must create a double bond to complete the octet of
every atom, and when there is a choice of where to form the
double bond, resonance structures occur. For instance, in the
dot structure for the NO_3^- ion above, the double bond could be
placed in any one of three places when you construct the dot
formula. Therefore, three resonance structures occur. The
actual structure of NO_3^- is a sort of average of these three.
Each bond is approximately 1-1/3 bonds.

Self-Test

14. Draw all resonance structures for

(a) SO_3

(b) N_3^- (structure, N······N······N)

(c) HCO_2^- (structure, H······C $\begin{smallmatrix} O \\ \\ O \end{smallmatrix}$)

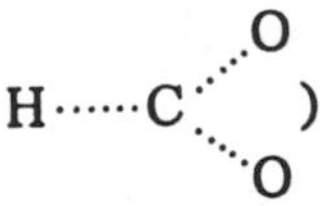

15. Arrange the following in order of predicted decreasing C—O bond length. Specify the average C—O bond order in each.

(a) $:\ddot{O}\!=\!C\!=\!\ddot{O}:$

(b) $:C\!\equiv\!O:$

(c) 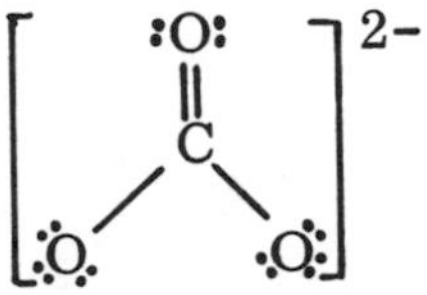 $\left[\begin{smallmatrix} :O: \\ \| \\ C \end{smallmatrix} \right]^{2-}$ (one of three resonance structures)

(d) 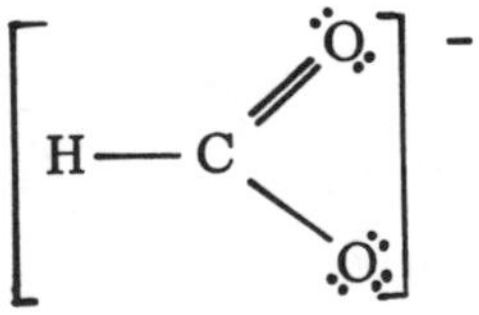 $\left[H\!-\!C \right]^{-}$ (one of two resonance structures)

(e) 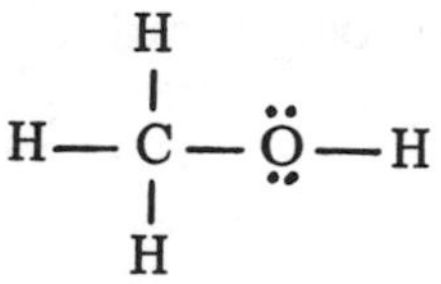
H—C—Ö—H with H above and below C

Answer _____

New Terms

resonance
resonance hybrid

4.7 COORDINATE COVALENT BONDS

Objectives

To see how atoms can use unshared electron pairs to form

additional bonds.

Review

The coordinate covalent bond is a bookkeeping device that is sometimes convenient to use in accounting for the bonding in some compounds. Remember that the properties of a covalent bond do not depend on the origin of the electrons shared between the atoms.

Self-Test

16. Draw electron-dot formulas for the following showing how the structure can be explained in terms of coordinate covalent bonding.

    (a) H_3O^+ (O is central atom)

    (b) BF_4^- (B is central atom)

New Terms

coordinate covalent bond   addition compound
dative bond

4.8   POLAR MOLECULES AND ELECTRONEGATIVITY

Objectives

To understand that in most molecules electrons are not shared equally between atoms because different atoms have different tendencies to attract electrons.

Review

Remember that electronegativity refers to the attraction an atom has for electrons in a bond, and that it is the difference in electronegativity that determines the polarity of a bond, as well as which end of a polar bond carries the negative charge. An important point to remember from this section is that bonds can vary anywhere between essentially 100% covalent to essentially 100% ionic.

Molecules having polar bonds can be nonpolar if the effects of the bond dipoles cancel. This will occur if the molecule has a symmetrical shape. (In Chapter 5 you will learn how to predict molecular shapes.)

You should know how electronegativity varies in the periodic table - increasing from left to right in a period and decreasing from top to bottom in a group. Keep in mind the margin figure on Page 129.

Self-Test

17.   Use Table 4.6 in the text to determine which end of the following bonds carries a partial negative charge.

(a) Sb—H _____   (c) C—O _____

(b) P—S _____   (d) Br—S _____

18.   <u>Without</u> referring to Table 4.6, but using a periodic table, predict which atom in each of the following sets is most electronegative.

(a) Ga, Bi, As, In _____

(b) Cl, P, Br, Bi _____

New Terms

electronegativity   polar
dipole   electropositive
dipole moment

4.9   OXIDATION AND REDUCTION, OXIDATION NUMBERS

Objectives

To learn that many chemical reactions can be explained in terms of the transfer of electrons, either partially or completely, from one atom to another. You should become familiar with, and be able to use <u>oxidation numbers</u> to keep tabs on electrons.

Review

Be sure you know the meaning of the terms oxidation and reduction. You should be able to identify the oxidizing agent and reducing agent in a chemical reaction. The definitions of oxidizing agent and reducing agent are sometimes confusing. Just remember that if a substance is oxidized (i.e., loses electrons), we call it a reducing agent; a substance that is reduced is an oxidizing agent.

It is important to learn the rules for assigning oxidation numbers. Practice on the questions in the Self-Test below.

Self-Test

19. Assign oxidation numbers to the atoms in the following formulas.

(a) $KClO_3$ _____

(b) $MnO_4{}^{2-}$ _____

(c) $S_2O_3{}^{2-}$ _____

(d) $SiCl_4$ _____

(e) BrF_3 _____

(f) P_4O_6 _____

(g) $SbCl_6{}^-$ _____

(h) $C_{12}H_{22}O_{11}$ _____

(i) $S_3{}^{2-}$ _____

(j) O_3 _____

20. Identify the oxidizing and reducing agent in the following reactions.

(a) $H_2 + Cl_2 \longrightarrow 2HCl$ _____

(b) $2Na_2S_2O_3 + I_2 \longrightarrow 2NaI + Na_2S_4O_6$ _____

(c) $3O_2 + C_2H_4 \longrightarrow 2CO_2 + 2H_2O$ _____

(d) $K_2Cr_2O_7 + 14HCl \longrightarrow 3Cl_2 + 2KCl + 2CrCl_3 + 7H_2O$

_____

New Terms

oxidation   reducing agent
reduction   oxidation number
redox   oxidation state
oxidizing agent

4.10   THE NAMING OF CHEMICAL COMPOUNDS

Objectives

To be able to name simple inorganic compounds.

Review

The rules for naming inorganic compounds are given in this section. If you have had chemistry in high school, you should remember many of them. If you've never had chemistry before, it is important that you learn how to name simple inorganic compounds. Besides increasing your ability to follow what's going on in class, you will probably need to know the names of chemicals if you ever have to get them from the storeroom or order them from a catalog.

Self-Test

21.   Name the following:

(a) $NiCl_2$ _____

(b) $Cr_2(CO_3)_3$ _____

(c) $Sr(NO_3)_2$ _____

(d) K_2SO_4 _____

(e) SF_6 _____

(f) CCl_4 _____

22.   Write formulas for the following:

(a)   tin(IV) oxide   _____

(b) boron trichloride _____

(c) calcium bicarbonate _____

(d) iron(III) oxide _____

(e) dinitrogen pentoxide _____

(f) sodium phosphide _____

23. What is the formula of the sodium salt of periodic acid?

_____

24. What is the name of the acid H_2CrO_4? _____

25. What would be the name for the acid H_2Se? _____

26. Arsenic acid has the formula, H_3AsO_4.

(a) What is the formula of arsenous acid? _____

(b) What is the formula for sodium arsenite? _____

(c) What is the name of KH_2AsO_4? _____

New Terms

inorganic compound
organic compound
trivial name
binary compound
Stock system

binary acid
oxoacids
neutralization
salt
acid salt

Answers to Self-Test Questions

1. (a) $\cdot\overset{\cdot}{\underset{\cdot}{Si}}\cdot$ (b) $:\overset{\cdot\cdot}{\underset{\cdot\cdot}{Cl}}\cdot$ (c) $\cdot Sr\cdot$ (d) $:\overset{\cdot}{As}\cdot$

2. (a) $K\times + \cdot\overset{\cdot\cdot}{\underset{\cdot\cdot}{Cl}}: \longrightarrow K^+$, $\left[\times\overset{\cdot\cdot}{\underset{\cdot\cdot}{Cl}}:\right]^-$ (b) $Na\times$ $+ \cdot\overset{\cdot}{\underset{\cdot}{O}}: \longrightarrow 2Na^+$, $\left[\overset{\cdot\times}{\underset{\cdot\cdot}{\times O}}:\right]^{2-}$
    $Na\times$

3. $3s^2 3p^6 3d^{10}$ 4.(a) $Cr_3(PO_4)_2$ (b) $CuSO_4$ (c) $Ga(ClO_2)_3$
(d) $Cr_2(C_2O_4)_3$ (e) $(NH_4)_2S$ 5.(a) phosphate ion   (b) sulfate
ion   (c) chlorite ion   (d) oxalate ion   (e) ammonium ion
6. law of conservation of energy   7. The lattice energy of $Li^{2+}F^{2-}$
isn't large enough to compensate for the very large endothermic

second ionization energy of Li and the energy needed to place a second electron into F^-.   8. energy

9.

$$H : \overset{\overset{\times\bullet}{}}{\underset{\underset{\times\bullet}{}}{Si}} : H \qquad H : \overset{\overset{\times\bullet}{}}{\underset{\underset{\bullet\bullet}{}}{P}} : H \qquad H : \overset{\overset{\bullet\bullet}{}}{\underset{\underset{\bullet\bullet}{}}{S}} : H \qquad H : \overset{\bullet\bullet}{\underset{\bullet\bullet}{Cl}} :$$

with H above Si and above P.

10. (a)

$$\ddot{C}l - \underset{\underset{:\ddot{C}l:}{|}}{\overset{\overset{:\ddot{C}l:}{|}}{Si}} - \ddot{C}l:$$

(b) $:\ddot{C}l - \ddot{F}:$ with $:\ddot{F}.$ and $.\ddot{F}:$

(c) $:\ddot{C}l - \ddot{S} - \ddot{C}l:$

(d) $\left[\underset{\underset{:\ddot{O}:}{|}}{\overset{\overset{:\ddot{O}:}{||}}{:\ddot{O} - S - \ddot{O}:}} \right]^{2-}$

(e) $:C \equiv O:$

(f) $\left[:\ddot{O} - \ddot{C}l - \ddot{O}: \right]^-$

11. It must be approximately 3.   12.(a) I (b) II   (c) II
(d) approximately 1.31 Å   13. The vibrational frequencies of bonds.

14. (a)

$$\overset{:O:}{\underset{:\ddot{O}. \quad .\ddot{O}:}{\overset{||}{S}}} \longleftrightarrow \overset{:\ddot{O}:}{\underset{.\ddot{O}. \quad .\ddot{O}:}{\overset{|}{S}}} \longleftrightarrow \overset{:\ddot{O}:}{\underset{:\ddot{O}. \quad .\ddot{O}.}{\overset{/ \quad \diagdown}{S}}}$$

(b) $:N \equiv N - \ddot{N}: \leftrightarrow :\ddot{N} - N \equiv N: \leftrightarrow :\ddot{N} = N = \ddot{N}:$

(c)

$$H - C \overset{\overset{.\ddot{O}.}{\diagup\diagup}}{\underset{.\ddot{O}:}{\diagdown}} \longleftrightarrow H - C \overset{\overset{.\ddot{O}:}{\diagup}}{\underset{.\ddot{O}.}{\diagdown\diagdown}}$$

15. e > c > d > a > b

16.(a) $H \overset{\times}{.} \overset{\overset{\bullet\bullet}{}}{\underset{\underset{H}{\times\bullet}}{O}} \overset{\bullet}{.} H$

(b) $\left[:\overset{\overset{:\ddot{F}:}{\times\bullet}}{\underset{\underset{:\ddot{F}:}{\times\bullet}}{F}} \overset{\times}{.} B \overset{\bullet\bullet}{.} \ddot{F}: \right]^-$

17.(a) H (b) S (c) O
(d) Br

18.(a) As (b) Cl 19.(a) K, 1+; Cl, 5+; O, 2− (b) Mn, 7+; O, 2− (c) S, 2+; O, 2− (d) Si, 4+; Cl, 1− (e) Br, 3+; F, 1− (f) P, 3+; O, 2− (g) Sb, 5+; Cl, 1− (h) C, zero; H, 1+; O, 2− (i) S, −2/3 (j) O, zero   20. Oxidizing agents: (a) Cl_2 (b) I_2 (c) O_2 (d) $K_2Cr_2O_7$ Reducing agents: (a) H_2 (b) $Na_2S_2O_3$ (c) C_2H_4 (d) HCl   21.(a) nickel(II) chloride (b) chromium(III) carbonate (c) strontium nitrate (d) potassium sulfate (e) sulfur(VI) fluoride or sulfur hexafluoride (preferred)

(f) carbon(IV) chloride or carbon tetrachloride (preferred)
22.(a) SnO_2 (b) BCl_3 (c) $Ca(HCO_3)_2$ (d) Fe_2O_3 (e) N_2O_5
(f) Na_3P 23. $NaIO_4$ 24. chromic acid   25. hydroselenic acid
26.(a) H_3AsO_3 (b) Na_3AsO_3 (c) potassium dihydrogen arsenate

5 COVALENT BONDING AND MOLECULAR STRUCTURE

This chapter describes the modern theories of chemical bonding and how they can account for (or predict, in some cases) molecular structure. The thing to keep in mind throughout discussions of the different approaches is that each theory is attempting to describe the same thing and each, in its own way, succeeds to a degree. The theories, then, present alternative views of the same phenomenon.

5.1 MOLECULAR SHAPES

Objectives

To learn to recognize the five principle shapes that can be used to describe most molecular structures. You should learn the geometric properties of these five structures.

Review

Study the five basic geometric shapes described in this section. You should know the bond angles characteristic to each molecular shape, and you should be able to sketch the shapes.

Self-Test

1. On a separate sheet of paper, sketch the shapes of a tetrahedron, a trigonal bipyramid, and an octahedron. Check yourself by comparing your drawings with those in the text

on Pages 141 and 142.   Practice until your drawings are reasonably accurate.

2.   What are the bond angles in

   (a)   a linear molecule?   _____

   (b)   a tetrahedral molecule?   _____

   (c)   an octahedral molecule?   _____

   (d)   a trigonal bipyramidal molecule?   _____

   (e)   a planar triangular molecule?   _____

New Terms

linear molecule   trigonal bipyramid
planar triangular molecule   octahedron
tetrahedron

5.2   VALENCE SHELL ELECTRON-PAIR REPULSION THEORY

Objectives

To learn how to predict the shape of a molecule from the molecule's Lewis structure.

Review

The key to success in applying this very simple theory is the ability to construct Lewis structures for molecules and ions. If necessary, review the procedure for drawing Lewis structures on Pages 119 to 122.

The principle behind the VSEPR theory is that electron pairs in the valence shell of an atom tend to get as far apart as possible so that repulsions between them are a minimum.   The arrangements of electron pairs that give minimum repulsions are given in Figure 5.1.   You should recognize these as the five shapes discussed in the previous section.

When there are lone pairs (unshared pairs) of electrons in the valence shell of the central atom they influence molecular shape.   In describing molecular shapes, remember that it is the arrangement of atoms that is specified and not how the electron pairs are arranged about the central atom. In SO_2, for example (Page 145), the electrons are arranged at the corners of a planar triangle. (Note that a multiple bond behaves just like a single bond for the purposes of predicting molecular shapes by the VSEPR theory.) One of these corners is occupied by a lone pair and the other two corners are occupied by oxygen atoms. In describing the shape of SO_2 we state how the two oxygens and the sulfur are arranged, and we ignore the lone pair. Thus SO_2 is said to be nonlinear or bent. It is not said to be planar triangular!

Study the shapes found for differing numbers of electron pairs and differing numbers of lone pairs in Figures 5.2, 5.3, 5.4, and 5.5. Be sure you know the names corresponding to the various structures. Remember that in the trigonal bipyramidal arrangement, the lone pairs are always found in the central triangular plane. The various shapes are summarized in Table 5.1. Although it is best if you can draw the shapes and deduce the structures from your drawings, you can also obtain correct answers if you know the contents of Table 5.1 thoroughly.

Self-Test

3. For each of the following, predict the arrangement of electron pairs and the molecular structure.

	electron arrangement	molecular structure
(a) AsH_3		
(b) $AlCl_4^-$		
(c) BrF_3		
(d) NO_2^-		
(e) ICl_2^-		
(f) H_3O^+		

New Terms

electron pair repulsion theory
lone pair

5.3   VALENCE BOND THEORY

Objectives

To describe how this theory views the formation of a chemical bond. You should learn the basic ideas on which the theory is based.

Review

This theory says that a bond is formed by sharing a pair of electrons between overlapping atomic orbitals. Only two electrons can be shared between two orbitals. When orbitals come together, they must each have one electron, or one must be empty if the other orbital supplies two electrons (a coordinate covalent bond). When p orbitals are used in bonding, bond angles tend toward 90°.

Self-Test

4. What would you predict for the structure (shape and bond angles) of:

(a) H_2S _____

(b) PH_3 _____

New Terms

valence bond theory
overlap of orbitals

5.4   HYBRID ORBITALS

Objectives

To see how the atomic orbitals of an atom can mix to form hybrid orbitals that possess new directional properties. You should learn how molecular shapes can be accounted for by the use of hybrid orbitals. You should learn the directional properties of the different kinds of hybrid orbitals, and how the VSEPR theory can be used to predict which kinds of hybrid orbitals an atom will use to form bonds.

Review

Learn the orientations (summarized in Figure 5.12) of the different hybrid orbital sets in Table 5.2. In applying the information in this table to describing molecular structure, remember that when a central atom enters into bonding with several others, it must supply one unpaired electron for each of the atoms to which it is bonding. Review the descriptions for BeH_2, CH_4 and SF_6. Note that additional electron pairs belonging to the central atom which are not used in bonding can reside in hybrid orbitals, too. This is described for H_2O and NH_3. Remember that when hybrids are formed, s and p orbitals are used first, followed by d orbitals.

For most molecules, the VSEPR thoery complements the valence bond theory well because it allows us to predict the orientations of the electron pairs around an atom. This information, in turn, tells us which kind of hybrid orbitals the atom uses. Review Example 5.4 in the text before taking the following Self-Test.

Self-Test

5. What kind of hybrid orbitals are used by the central atom in each of the following? What molecular structure is expected?

    (a) $SiCl_4$ _____   _____

    (b) BCl_3 _____   _____

    (c) XeF_4 _____   _____

    (d) PCl_5 _____   _____

    (e) $SnCl_2$ _____   _____

6. The ion $SbCl_6^-$ is formed by the attachment of a Cl^- ion to an $SbCl_5$ molecule. This involves the creation of a coordinate covalent bond. Illustrate the bonding in $SbCl_6^-$ by means of an orbital diagram. Use different symbols for the Sb and Cl electrons.

New Terms

hybrid orbitals

5.5 MULTIPLE BONDS

Objectives

To see how multiple bonds are described in terms of
orbital overlap.

Review

Multiple bonds are usually formed when unpaired electrons
would occur on atoms if the rules in the last section were followed. In ethylene, for example, each carbon atom is bonded to
three other atoms.

$$ \underset{H}{\overset{H}{\diagdown}} C \text{—} C \underset{H}{\overset{H}{\diagup}} $$

This requires three unpaired electrons on the carbon atoms, and
if no double bond were formed, we would have the situation,

C $\underline{\uparrow}$ $\underline{\uparrow x}$ $\underline{\uparrow x}$ $\underline{\uparrow x}$ (x's are electrons
     sp^3 from other atoms)

This leaves an unpaired electron in an sp^3 hybrid orbital. This
tends to displease Mother Nature. As a result, in this kind of
situation we only use three hybrid orbitals (sp^2) for σ-bonds.
The remaining electron remains in an unhybridized orbital and
can then pair with another electron on the neighboring carbon
atom to form a π-bond.

Lewis structures are based on the valence bond description
of bonding. If a Lewis structure has multiple bonds, you should
account for the bonding as described in this section. Be sure
you understand the meaning of σ-bond and π-bond. Remember
that the basic molecular framework or skeleton of a molecule is
accounted for by σ-bonds, which determine the kinds of hybrid
orbitals the atoms use. Any unhybridized orbitals containing unpaired electrons then participate in π bonding.

Self-Test

7. Indicate the number of σ- and π-bonds in the bonds of the following molecules:

    (a) CO_2 _____

    (b) CO_3^{2-} _____

    (c) N_2 _____

    (d) C_2^{2-} _____

    (e) GeH_4 _____

8. What kinds of hybrid orbitals are used by the atoms in the molecules below, and what kinds of bonds (σ, π) exist between the atoms?

    (a)

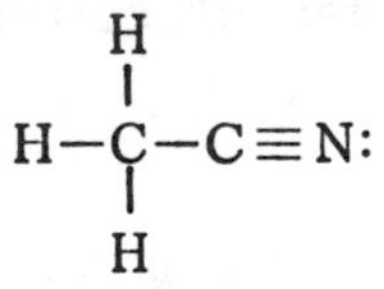

$$\begin{array}{c} H \\ | \\ H-C-C\equiv N: \\ | \\ H \end{array}$$

    (b)

$$\begin{array}{c} H \quad H \qquad H \\ | \quad\; | \quad\;\; | \\ H-C-C=\overset{..}{N}-\underset{..}{N}-H \\ | \\ H \end{array}$$

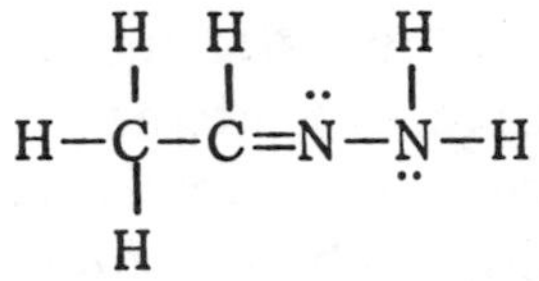

New Terms

σ-bond
π-bond

5.6 RESONANCE

Objectives

To relate the Lewis structures for molecules that exhibit resonance to the valence bond description of bonding.

Review

The point that is made here is that Lewis structures are, in effect, simplified versions of valence bond structures.

New Terms

5.7 MOLECULAR ORBITAL THEORY

Objectives

> To understand the concept of how molecular orbitals are created from atomic orbitals. You should learn the difference between bonding and antibonding molecular orbitals and how the electronic structure of a molecule is obtained by filling molecular orbitals. You should also learn how molecular orbital theory avoids the idea of resonance.

Review

Molecular orbitals appear to be formed by constructive and destructive interference of the electron waves of atomic orbitals. We obtained their descriptions by alternately adding and subtracting atomic orbitals that overlap from different atoms. Bonding orbitals concentrate electron density between nuclei, thereby binding the atoms together. Antibonding orbitals place electron density outside the region between nuclei and, when occupied by electrons, decrease the stability of the molecule.

Molecular orbitals are filled following the same rules for filling atomic orbitals in atoms: (1) electrons go into lowest energy orbitals first; (2) no more than two electrons may occupy the same orbital; (3) when there are orbitals of the same energy, the electrons spread out as much as possible with spins in the same direction.

Learn how to calculate the net bond order from the number of bonding and antibonding electrons. You should also learn the energy level diagram for diatomic molecules (Figure 5.22).

Notice that molecular orbital theory avoids resonance by permitting molecular orbitals to spread out over more than two atoms.

Self-Test

9. On a separate piece of paper, sketch the molecular orbital
    energy level diagram for diatomic molecules having the
    second shell (2s and 2p subshells) as their valence shell.
    Use this diagram to answer the following questions.

    (a) Is the molecule Be_2 stable?   _____

    (b) Which species is more stable, O_2 or O_2^+?   _____

    (c) Which species is more stable, N_2 or N_2^+?   _____

    (d) Which has the shorter bond, C_2^- or C_2^{2-}?   _____

New Terms

molecular orbital   net bond order
bonding orbital   delocalized molecular orbital
antibonding orbital

Answers to Self-Test Questions

1. See Page 141, 142 of text.   2.(a) 180°   (b) 109.5°   (c) 90°
(d) 120° (in triangular plane), 90° (between triangular plane and
vertical bond)   (e) 120°

3.   electron arrangement   molecular structure
(a)   tetrahedral   pyramidal
(b)   tetrahedral   tetrahedral
(c)   trigonal bipyramidal   T-shaped
(d)   planar triangular   angular
(e)   trigonal bipyramidal   linear
(f)   tetrahedral   pyramidal

4.(a) angular (bent) molecule, H—S—H angle = 90° (b) pyramidal
molecule, H—P—H angles = 90°

5.(a) sp^3, tetrahedral   (b) sp^2, planar triangular   (c) sp^3d^2,
square planar   (d) sp^3d, trigonal bipyramidal   (e) sp^2, angular
(bent)

6. Sb $\underline{\uparrow x}$ $\underline{\uparrow x}$ $\underline{\uparrow x}$ $\underline{\uparrow x}$ $\underline{\uparrow x}$ $\underline{xx}$ __ __ __   (x = Cl electrons)

$\underbrace{\qquad\qquad}_{sp^3d^2}$ $\nearrow$ unhybridized 5d orbitals

$\Big\lfloor$ coordinate covalent bond

7.(a)   two σ-bonds, two π-bonds
  (b)   three σ-bonds, one π-bond
  (c)   one σ-bond, two π-bonds
  (d)   one σ-bond, two π-bonds
  (e)   four σ-bonds

8.(a)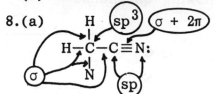

  (b)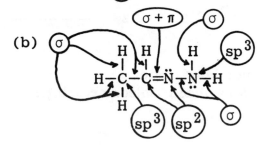

9.(a)   no; net bond order = 0
  (b) O_2^+ is more stable since net bond order is greater than
         for O_2.
  (c) N_2, larger net bond order
  (d) C_2^{2-} (same structure as N_2), larger net bond order.

6 CHEMICAL REACTIONS IN AQUEOUS SOLUTION

Most chemical reactions are carried out in solution and very often the solvent is water. This chapter focuses on a variety of aspects dealing with reactions in water solutions. The chemistry described here is very important and must be learned thoroughly. Once again, if you have had high school chemistry, you may already know much of the material in this chapter. Review it anyway and test yourself to be sure. If you've not had high school chemistry, be sure to master this material well. You should be able to answer the self-test questions without having to look at rules or tables, unless directed to do so in the question.

6.1 SOLUTION TERMINOLOGY

Objectives

To become familiar with terms commonly used to describe solutions and their components.

Review

Be sure you understand and are familiar with the terms introduced in this section so that you can follow discussions when they are used later on.

1. Sugar has a solubility of 211 g per 100 g of water at 25°C.
    A solution containing 215 g of sugar in 100 g of water at
    25°C would be described by which term (or terms)?
    (a) concentrated   (b) dilute   (c) unsaturated
    (d) supersaturated (e) saturated   _____

2. In the solution in Question 1, the solute is _____
    and the solvent is   .
    _____

New Terms

solvent   solubility
solute   saturated
concentrated   unsaturated
dilute   supersaturated

6.2   ELECTROLYTES

Objectives

To understand what takes place when an ionic solid dissolves in water. You will see that these substances break
apart in water, as do certain molecular substances, to
produce ions that are free to roam about.

Review

Remember that essentially all ionic solids are virtually 100%
dissociated in aqueous solution. The ions are surrounded by
water molecules and are said to be hydrated. Some covalent substances also form ions in aqueous solution by reaction with the
solvent. An important ion produced in this way is the hydronium
ion, H_3O^+. Substances that are 100% dissociated are strong electrolytes; those that are partially dissociated are weak electrolytes,
and those that are undissociated are nonelectrolytes.

Self-Test

3. Potassium nitrate is an ionic solid. Write an equation repre-
    senting its dissociation in water.

4. What is the name of the hydrated proton?_____

New Terms

dissociation   strong electrolyte
electrolyte   weak electrolyte
hydronium ion   nonelectrolyte
hydrated ion

6.3   CHEMICAL EQUILIBRIUM

Objectives

To learn that chemical reactions are able to proceed in two
directions, from reactants to products and from products
to reactants. You should become familiar with the concept
of dynamic equilibrium.

Review

The concept of dynamic equilibrium in chemistry is extremely important. Remember that chemical reactions are generally able
to proceed in both forward and reverse directions. When opposing reactions are occurring at the same speed there is no change
in the amounts of reactants or products. We use the term
position of equilibrium to describe how far toward completion the
reaction proceeds before equilibrium is reached. In a reaction

$$A \rightleftharpoons B$$

the position of equilibrium lies far to the right if a lot of B is
produced from A by the time equilibrium is attained.

Self-Test

5. Write chemical equations showing the equilibria in the dissociation of the following weak electrolytes (see Table 6.1):

(a) NH_3 _____

(b) HCN   _____

New Terms

dynamic equilibrium
position of equilibrium

6.4   IONIC REACTIONS

Objectives

To describe ionic reactions in solution by chemical equations that depict reactions between ions. You should also
learn how to predict the products of an ionic reaction on
the basis of solubility or the possible formation of an undissociated product or a gas. You should learn the solubility rules given here.

Review

Remember that the ionic equation for a reaction is obtained
by writing the formulas of all soluble ionic compounds in dissociated form. Insoluble compounds are written in undissociated
form (the list of solubility rules in this section should be
memorized so that it should not be necessary to turn to them
constantly). An example is the reaction of sodium sulfate with
barium nitrate.

$$Na_2SO_4 + Ba(NO_3)_2 \longrightarrow 2NaNO_3 + BaSO_4$$

This type of "exchange of partners" reaction is called metathesis.
It is also often called a double replacement reaction. In writing
the reaction, be sure you have the correct formulas of the products; then proceed to balance the equation. In this example
three of the compounds are soluble (Na_2SO_4, $Ba(NO_3)_2$ and
$NaNO_3$) while only one is insoluble ($BaSO_4$). Writing the soluble
ones in dissociated from gives the ionic equation,

$$2Na^+ + SO_4^{2-} + Ba^{2+} + 2NO_3^- \longrightarrow 2Na^+ + 2NO_3^- + BaSO_4(s)$$

The net ionic equation is obtained by eliminating ions that appear
the same on both sides of the arrow.

$$Ba^{2+} + SO_4^{2-} \longrightarrow BaSO_4(s)$$

An ionic reaction will proceed to completion if a net ionic equation can be written without all of the ions cancelling. For example, nothing is observed to happen when solutions of $CaCl_2$ and $Ba(NO_3)_2$ are mixed. The "reaction,"

$$CaCl_2 + Ba(NO_3)_2 \longrightarrow BaCl_2 + Ca(NO_3)_2$$

does not occur because all the reactants and products are soluble.

$$Ca^{2+} + 2Cl^- + Ba^{2+} + 2NO_3^- \longrightarrow Ba^{2+} + 2Cl^- + Ca^{2+} + 2NO_3^-$$

Notice that all ions cancel. Nothing is left, so there is no reaction.

Remember that all ions won't cancel if:

    (a) a product is insoluble.
    (b) a product is a weak electrolyte.
    (c) a product is a gas.

Substances that fit the last two conditions are found in Tables 6.1, 6.2 and 6.3 of the text.

Sometimes a reactant will also be a solid or weak electrolyte. A reaction will occur in these instances only if a product is less soluble, a weaker electrolyte, or a gas. You are not expected to predict the outcome of these reactions now, since you haven't been given rules about relative solubilities or degrees of dissociation.

Self-Test

6. Study the solubility rules before answering this question. Use this question to test your knowledge of the rules. If the substance is soluble, write S; if it is insoluble, write I.

    (a) KNO_3 _____   (e) ZnO _____

    (b) $MgSO_4$ _____   (f) $PbSO_4$ _____

    (c) AgI _____   (g) $Ni(OH)_2$ _____

    (d) $FeCO_3$ _____   (h) $(NH_4)_2CrO_4$ _____

7. Write ionic and net ionic equations for these reactions:

    (a) $NaI + AgNO_3 \longrightarrow AgI + NaNO_3$

    (b) $Pb(NO_3)_2 + Ba(OH)_2 \longrightarrow Pb(OH)_2 + Ba(NO_3)_2$

(c) $AgCl + NaBr \longrightarrow AgBr + NaCl$

(d) $ZnCl_2 + Na_2CO_3 \longrightarrow 2NaCl + ZnCO_3$

(e) $BaCl_2 + 2NaBr \longrightarrow BaBr_2 + 2NaCl$

8.   Write ionic and net ionic equations for these reactions:

(a) $CaCO_3 + 2HCl \longrightarrow H_2O + CO_2 + CaCl_2$

(b) $(NH_4)_2SO_4 + 2NaOH \longrightarrow 2NH_3 + 2H_2O + Na_2SO_4$

(c) $Na_2C_2O_4 + 2HCl \longrightarrow 2NaCl + H_2C_2O_4$

(d) $BaCO_3 + H_2SO_4 \longrightarrow BaSO_4 + H_2O + CO_2$

(e) $K_2SO_3 + H_2SO_4 \longrightarrow K_2SO_4 + H_2O + SO_2$

9.   Write molecular, ionic, and net ionic equations for the reaction, if any, that would occur between the following:

(a) $NiCl_2$ and Na_2CO_3 (d) Na_2SO_4 and $Mg(NO_3)_2$

(b) $NaCl$ and $CaBr_2$ (e) Na_2S and HCl

(c) $MgCl_2$ and $NaOH$

New Terms

filtrate	ionic equation
precipitate	net ionic equation
metathesis	spectator ion
double replacement	salts
molecular equation	

6.5   ACIDS AND BASES IN AQUEOUS SOLUTION

Objectives

To learn how acids and bases are defined and how acids and bases react with each other.

Review

Remember:

acids produce H_3O^+ when dissolved in water
bases produce OH^- when dissolved in water

Many acids contain hydrogen that attaches itself as H^+ to water molecules to produce H_3O^+. In many cases H_3O^+ is simply indicated as H^+ since this is the "active ingredient" in the H_3O^+ ion during chemical reactions. Metal hydroxides, when soluble, dissociate to give OH^- in solution.

Acids and bases react with each other to produce a salt and water. A salt is a general term that is used to refer to any ionic compound except metal hydroxides, which are called bases. An example of an acid-base neutralization reaction is

$$2HCl + Ba(OH)_2 \longrightarrow BaCl_2 + 2H_2O$$

Polyprotic acids (those able to supply more than one H^+ per molecule of acid) can be partially neutralized to give acid salts; for example, $NaHSO_4$ from the partial neutralization of H_2SO_4. Bases generally do not undergo partial neutralizations; instead, all of the OH^- of a base is neutralized in the formation of salts which can be isolated from solution.

You should also remember that metal oxides are basic. When they dissolve in water they produce hydroxides because of the reaction,

$$O^{2-} + H_2O \longrightarrow 2OH^-$$

Even if they are insoluble, metal oxides still react with acids; for example,

$$Fe_2O_3 + 6H^+ \longrightarrow 2Fe^{3+} + 3H_2O$$
(rust)

Nonmetal oxides are acidic. Many give acids when they dissolve. For example,

$$CO_2 + H_2O \longrightarrow H_2CO_3$$

Self-Test

10. Write balanced molecular equations for the neutralization reaction between:

    (a) $Ca(OH)_2$ and HCl

    (b) NH_3 and H_2SO_4 (complete neutralization)

    (c) $Al(OH)_3$ and H_2SO_4 (complete neutralization)

    (d) NaOH and HNO_3

    (e) MgO and H_2SO_4 (complete neutralization)

11. Write formulas for all salts formed between $Ca(OH)_2$ and

    (a) $H_2C_2O_4$ _____

    (b) H_2CO_3 _____

    (c) H_3AsO_4 _____

    (d) HBr _____

12. Indicate whether water solutions of the following would be expected to be acidic or basic.

    (a) SO_3 _____   (c) P_4O_{10} _____

    (b) BaO _____   (d) SeO_3 _____

New Terms

acid	neutralization
base	monoprotic acid
strong acid	diprotic acid
weak acid	triprotic acid
strong base	polyprotic acid
weak base	acid salt

6.6   THE PREPARATION OF INORGANIC SALTS BY METATHESIS REACTIONS

Objectives

To learn methods that can be used to prepare salts by reactions in aqueous solution.

Review

The techniques discussed in this section call upon what you have learned in Sections 6.4 and 6.5. Your ability to choose a method to prepare a salt obviously depends on how well you've learned this previous material. If you have difficulty with most of the problems in the Self-Test, go back and review Sections 6.4 and 6.5. In particular, pay attention to solubility rules and the factors that cause reactions to go to completion.

Precipitation Reactions. These make use of the formation of a precipitate to obtain the desired product. You need to remember the solubility rules to apply this method. Remember to use soluble reactants that give only one insoluble product.

Example 6.1

How can we prepare $CuCO_3$ by a precipitation reaction?

Solution

The product, $CuCO_3$, is insoluble; therefore we want the other product to be soluble. Also, we want to begin with a soluble carbonate and a soluble copper salt. Possible reactants are $CuCl_2$ and Na_2CO_3. Both are soluble and the choice of the sodium salt to provide the CO_3^{2-} ensures us that the other product in the metathesis reaction will be soluble.

$$CuCl_2 + Na_2CO_3 \longrightarrow 2NaCl + CuCO_3$$
$$\text{(sol.)} \quad \text{(sol.)} \quad \quad \text{(sol.)} \quad \text{(insol.)}$$

Neutralization Reactions. The desired salt is derived from the cation of a base and the anion of an acid. For instance, to prepare $CuBr_2$ you could use the reaction between $Cu(OH)_2$ and HBr,

$$Cu(OH)_2(s) + 2HBr(aq) \longrightarrow CuBr_2(aq) + 2H_2O$$

Since $Cu(OH)_2$ is insoluble, an excess of $Cu(OH)_2$ is used so that all of the HBr in solution is used up. Excess insoluble $Cu(OH)_2$ is removed by filtration and the solution that passes through the filter contains only $CuBr_2$ which can be recovered by evaporation.

This method is good because most metal hydroxides are insoluble and can be prepared from other readily available salts by reaction with a base. For instance,

$$Cu(NO_3)_2(aq) + 2NaOH(aq) \longrightarrow Cu(OH)_2(s) + 2NaNO_3(aq)$$

Reactions in Which a Product is a Gas. Reactions of metal carbonates were discussed in the text. Any metathesis reaction in which one product is a gas will leave the other product by itself in solution. Some sample reactions are:

$$K_2SO_3 + 2HClO_4 \longrightarrow 2KClO_4 + H_2O + SO_3(g)$$

$$FeS + 2HBr \longrightarrow FeBr_2 + H_2S(g)$$

Review also Examples 6.5 to 6.8 in the text before beginning the Self-Test.

Self-Test

13. How would you prepare the following by a precipitation reaction?

(a) $Fe(NO_3)_3$ (b) $AgBr$ (c) $BaBr_2$ (d) $NaOH$

14. How would you prepare the following by a neutralization reaction?

(a) $Cu(HSO_4)_2$ from $CuCl_2$ (c) $Mg(NO_3)_2$ from $MgCl_2$

(b) $NiSO_4$ from $NiCl_2$ (d) $Ca(NO_3)_2$ from CaO

15. How would you prepare the following by a reaction that produces a gas?

(a) $FeCl_2$ from $FeCO_3$ (d) $ZnCl_2$ from ZnS

(b) $Co(NO_3)_2$ from $CoCO_3$ (e) $Ca(NO_3)_2$ from $CaCl_2$

(c) $NaNO_2$ from NH_4NO_2

New Terms

6.7   OXIDATION-REDUCTION REACTIONS

Objectives

To learn to balance oxidation-reduction reactions by making use of the changes of oxidation numbers.

Review

This section and the next discuss alternative ways of balancing redox reactions. The approach taken by the oxidation-number-change (ONC) method and the ion-electron (IE) method (Section 6.8) are different in several respects. It is only necessary to assign oxidation numbers in the ONC method. The IE method does not employ oxidation numbers even though the same end result is achieved. The key to both methods is making the number of electrons gained equal to the number lost.

In the ONC method, be sure to calculate the number of electrons transferred <u>per formula unit</u> for the reactants. Place coefficients in the equation to make the total electron loss equal to the total electron gain. Balance the remainder of the equation by inspection.

Remember, there is never any reason to be unsure that an equation is balanced correctly. You can always count up the numbers of each kind of atom on each side of the arrow. Also remember that an equation is not balanced unless there is the same net charge on each side.

Self-Test

16. Balance the following by the oxidation-number-change method.

   (a) $PH_3 + N_2O \longrightarrow H_3PO_4 + N_2$

   (b) $NaIO_3 + Na_2SO_3 \longrightarrow Na_2SO_4 + NaI$

   (c) $PbO_2 + HCl \longrightarrow PbCl_2 + H_2O + Cl_2$

New Terms

oxidation-number-change method

6.8   BALANCING REDOX EQUATIONS BY THE ION–ELECTRON METHOD

Objectives

To learn a method of developing balanced net ionic equations for redox reactions in aqueous solution.

Review

Balancing a redox equation by this method is very simple if you remember to be sure to follow these steps:

1. divide the reaction into two half-reactions.

2. balance atoms other than H and O.

3. balance O and H (O first, then H) using H_2O and H^+, respectively, for reactions in acid solution.

4. count up the net charge on both sides of each half-reaction.

5. for each half-reaction, add electrons to the most positive (least negative) side to make the net charge on both sides of the arrow the same.

6. multiply the half-reactions by appropriate factors to make electron gain equal electron loss.

7. add the half-reactions.

8. cancel anything that appears the same on both sides of the equation.

To balance an equation for a reaction taking place in basic solution, first balance it as if it were taking place in acidic solution. Then follow the three-step procedure on Page 192 to convert the equation to basic solution. Remember to add the necessary number of OH^- to <u>both</u> sides of the equation - otherwise you will upset the balance.

When you use the ion-electron method it is extremely important to write the appropriate charges on each formula. If you mean H^+ and write H, without giving the charge, you will almost certainly get the number of electrons wrong. This, of course, will mean that you will multiply the half-reactions by the wrong factors and hence obtain an improperly balanced equation.

Self-Test

17. Balance the following by the ion-electron method. (All reactions in acid solution)

(a) $HNO_2 + I^- \longrightarrow I_2 + NO$

(b) $ClO_3^- + H_2S \longrightarrow Cl^- + S$

(c) $S_2O_8{}^{2-} + P \longrightarrow SO_4{}^{2-} + H_3PO_4$

18. Balance the following by the ion-electron method. (All reactions in basic solution)

   (a) $CrO_4{}^{2-} + SO_3{}^{2-} \longrightarrow CrO_2{}^- + SO_4{}^{2-}$

   (b) $HO_2{}^- + ClO_2 \longrightarrow ClO_2{}^- + O_2$

   (c) $MnO_4{}^- + NO_2{}^- \longrightarrow MnO_2 + NO_3{}^-$

   (d) $ClO^- + NH_3 \longrightarrow N_2H_4 + Cl^-$ (This reaction between bleach, OCl^-, and ammonia, NH_3, can produce poisonous hydrazine, N_2H_4. Be careful - don't mix household cleansers!)

New Terms

ion-electron method
half-reaction

6.9   QUANTITATIVE ASPECTS OF REACTIONS IN SOLUTION

Objectives

To express the concentration of solute in solution. You should learn how to solve problems dealing with the quantities of solutions used in chemical reactions.

Review

There are several ways of expressing concentration. Parts per hundred (percent) and parts per million (ppm) were discussed in the text. One of the most important concentration units is molarity, the ratio of moles of solute to liters of solution. If you know the number of moles of solute and the volume of the solution (in liters), the ratio gives molarity.

Example 6.2

What is the molarity of a solution containing 0.843 mol of NaCl in 750 ml of solution?

Solution

$$\text{molarity} = \frac{\text{moles}}{\text{liters}} = \frac{0.843 \text{ mol NaCl}}{0.750 \text{ liter solution}}$$

$$\text{molarity} = 1.12 \text{ M}$$

Molarity is a convenient conversion factor relating moles of solute to volume of solution. The molarity of the solution in Example 6.2 can be used to construct two conversion factors,

$$\frac{1.12 \text{ mol NaCl}}{1.00 \text{ liter}} \quad \text{and} \quad \frac{1.00 \text{ liter}}{1.12 \text{ mol NaCl}}$$

which can be used in calculations.

Example 6.3

What volume of 1.12 M NaCl contains 0.420 mol of NaCl?

Solution

Moles must cancel. Therefore,

$$0.420 \text{ mol NaCl} \times \left(\frac{1.00 \text{ liter}}{1.12 \text{ mol NaCl}}\right) = 0.375 \text{ liter}$$

or

$$0.420 \text{ mol NaCl} \times \left(\frac{1000 \text{ ml}}{1.12 \text{ mol NaCl}}\right) = 375 \text{ ml}$$

The answer is 375 ml of solution.

Example 6.4

How many moles of NaCl are in 250 ml of 1.12 M NaCl solution?

Solution

$$250 \text{ ml} = 0.250 \text{ liter}$$

$$0.250 \text{ liter} \times \left(\frac{1.12 \text{ mol NaCl}}{1.00 \text{ liter}}\right) = 0.280 \text{ mol NaCl}$$

We can also use milliliters directly,

$$250 \text{ ml} \times \left(\frac{1.12 \text{ mol NaCl}}{1000 \text{ ml}} \right) = 0.280 \text{ mol NaCl}$$

If you thoroughly understand the three problems above, as well as the example problems in the text, you are ready for the next Self-Test.

Self-Test

19. Calculate the molarity of the following solutions:

    (a) 1.14 mol KI in 1.50 liter of solution   _____

    (b) 0.240 mol $CaCl_2$ in 500 ml of solution   _____

    (c) 3.50 g of NaCl in 0.0500 liter of solution   _____

    (d) 4.25 g $MgSO_4$ in 75.0 ml of solution   _____

20. How many moles of $KClO_3$ are in 500 ml of 0.150 M solution?

    _____

21. How many moles of urea are in 250 ml of urine if the urea concentration is 0.320 M?

    _____

22. A normal adult excretes about 1500 ml of urine per day. If the urea concentration is 0.320 M, how many grams of urea are excreted per day? Urea has the formula, $CO(NH_2)_2$.

    _____

23. What volume of 0.250 M $CaCl_2$ are required to react completely with 300 ml of 0.150 M $AgNO_3$ according to the equation,
$$CaCl_2 + 2AgNO_3 \longrightarrow Ca(NO_3)_2 + 2AgCl$$

    _____

24. How many moles of solid AgCl will be formed if 300 ml of 0.240 M $AgNO_3$ are added to 200 ml of 0.480 M HCl? The reaction is
$$AgNO_3 + HCl \longrightarrow AgCl + HNO_3$$
    _____

New Terms

parts per million

6.10 EQUIVALENT WEIGHTS AND NORMALITY

Objectives

To define a quantity useful for dealing with the stoichiometry of acid-base and redox reactions and to extend this concept to reactions in solution.

Review

The equivalent weight (that is, the weight of one equivalent) is defined differently for redox reactions and for acid-base neutralization reactions.

<u>Acid-Base Reactions</u>. For an acid the equivalent weight is equal to the formula weight divided by the number of H^+ ions neutralized. If H_3PO_4 is completely neutralized to give PO_4^{3-}, the equivalent weight of H_3PO_4 is the formula weight divided by 3. On the other hand, if H_3PO_4 is only partially neutralized to give HPO_3^{2-}, only two of the three H^+ ions are neutralized and the equivalent weight is the formula weight divided by 2.

<u>Redox Reactions</u>. The equivalent weight of a substance is simply its mole weight divided by the number of electrons that it gains or loses in the reaction. Thus when $KMnO_4$ is reduced to MnO_2 the manganese undergoes a change of 3 electrons.

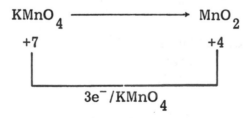

In this reaction the equivalent weight is the formula weight of $KMnO_4$ divided by 3.

$$\text{eq wt } KMnO_4 = \frac{\text{mole weight}}{3} = \frac{158 \text{ g}}{3} = 52.7 \text{ g}$$

$$1 \text{ eq } KMnO_4 = 52.7 \text{ g } KMnO_4$$

Note that there are 3 equivalents of $KMnO_4$ per mole in this reaction, or there is 1/3 of a mole per equivalent.

How about a reaction in which $K_2Cr_2O_7$ is converted to $CrCl_3$? What is the equivalent weight of $K_2Cr_2O_7$? In the reaction each Cr gains 3 electrons; since $K_2Cr_2O_7$ has two Cr atoms, each $K_2Cr_2O_7$ gains 6 electrons.

$$K_2Cr_2O_7 \longrightarrow 2CrCl_3$$

$$+6 \qquad\qquad\qquad +3$$

$$6e^-/K_2Cr_2O_7$$

The equivalent weight of $K_2Cr_2O_7$ is its formula weight divided by 6. There are six equivalents of $K_2Cr_2O_7$ per mole.

Remember that the whole rationale for equivalent weights is that <u>one equivalent of one substance reacts exactly with one equivalent of another.</u>

Redox
1 equivalent of oxidizing agent reacts with 1 equivalent of reducing agent.

Acid-Base
1 equivalent of acid reacts exactly with 1 equivalent of base.

<u>Normality</u>. Remember that normality is similar to molarity, except that the units are equivalents/liter. Also remember the useful relationship between normality (N) and molarity (M),

$$N = n \cdot M$$

where n is the number by which the formula weight must be divided to give equivalent weight (the number of electrons transferred in <u>redox</u>, or the number of H^+ or OH^- neutralized in acid-base reactions).

When solutions are reacted, remember that

$$N_A V_A = N_B V_B$$

where N_A and V_A are the normality and volume of solution A; N_B and V_B are the normality and volume of solution B.

Example 6.5

How many ml of 0.180 N $Ba(OH)_2$ are required to react completely with 22.0 ml of 0.0800 N HCl?

Solution

	acid	base
N	0.0800	0.180
V	22.0 ml	x

$$N_1 V_1 = N_2 V_2$$

$$(0.0800 \text{ N})(22.0 \text{ ml}) = (0.180 \text{ N})(x)$$

$$x = \frac{(0.0800 \text{ N})(22.0 \text{ ml})}{(0.180 \text{ N})} = 9.78 \text{ ml}$$

The volume of $Ba(OH)_2$ solution required is 9.78 ml.

Self-Test

25. Calculate the equivalent weight for the following substances. The products produced when they react are given in parentheses.

   (a) $NaBiO_3$ (Bi^{3+}) _____

   (b) $NaIO_3$ (I^-) _____

   (c) H_2SO_4 (SO_4^{2-}) _____

   (d) H_2SO_4 (HSO_4^-) _____

   (e) $Na_2S_2O_3$ $(S_4O_6^{2-})$ _____

26. $Na_2S_2O_4$ reacts with $CuSO_4$ in solutions containing ammonia to produce SO_3^{2-} and metallic copper. How many grams of $Na_2S_2O_4$ are needed to completely reduce 15.0 g of $CuSO_4$?

   _____

27. How many grams of K_2CrO_4 are needed to prepare 300 ml of 0.100 N solution if the solution will be used in a reaction in which the Cr is reduced to the +3 oxidation state?

   _____

28. How many ml of 0.100 N HCl react with 19.5 ml of 0.220 N KOH?

   _____

29. What is the normality of an H_2SO_4 solution if 27.0 ml of the solution is required to neutralize 14.3 ml of 0.35 N NaOH?

_____

New Terms

equivalent   normality
equivalent weight

6.11 CHEMICAL ANALYSIS

Objectives

To learn how the principles discussed in the previous sections of this chapter can be put to practical use for the purposes of chemical analysis.

Review

There are three important ideas developed in this section. The first of these involves analyzing a substance by forming and weighing a compound of known composition containing all of one component of the original sample. For example, if a mixture is known to contain silver, the compound AgCl can be formed by adding Cl^- to a solution of a weighed sample of the mixture. The insoluble AgCl is filtered, dried and weighed. From the known composition of AgCl, the weight of Ag in the precipitate can be calculated. Since the Ag came from the original sample, the weight of Ag in the sample is known. The practical application of this method, of course, depends on a knowledge of solubilities.

_____

Example 6.6

A 1.00-g sample of an ore known to contain silver was dissolved in HNO_3 and treated with Cl^-. 0.275 g of AgCl was obtained. What was the percent Ag in the ore?

Solution

From the weight of AgCl we calculate the weight of Ag. In one mole of AgCl (143.4 g) there is one mole of silver (107.9

g). Therefore,

$$0.275 \text{ g AgCl} \times \left(\frac{107.9 \text{ g Ag}}{143.4 \text{ g AgCl}} \right) = 0.207 \text{ g Ag}$$

(If you don't understand this calculation, review Section 2.3.)
All of this 0.207 g of Ag came from the ore. Therefore, the percent silver in the ore is

$$\%Ag = \frac{0.207 \text{ g}}{1.00 \text{ g}} \times 100 = 20.7\%$$

The second major point in this section involves the use of
the quantitative relationships of solution stoichiometry in a procedure called titration. One reactant in solution is added to another from a calibrated buret. If the concentration of the titrant
(the solution delivered from the buret) is known, the amount of
the reactant in the other solution can be calculated. Review Examples 6.23 and 6.24 in the text.

The third item discussed in this section is <u>dilution</u>. Remember the two simple relationships,

$$N_i V_i = N_f V_f$$

$$M_i V_i = M_f V_f$$

Self-Test

30. A 0.833-g sample of a mixture of NaCl and $CaCl_2$ was dissolved in water and treated with Na_2CO_3 to precipitate
$CaCO_3$. The precipitate was filtered and dried and found to
weigh 0.415 g. What percent of the original sample was
$CaCl_2$?

31. A 0.400-g sample of an alloy of iron and nickel was dissolved
in HCl to give Fe^{2+} and Ni^{2+}. The resulting solution was
titrated with 22.3 ml of 0.200 N $KMnO_4$, causing Fe^{2+} to be
oxidized to Fe^{3+}. What percent of the alloy is iron?

32. How many ml of 3.00 M HCl must be used to prepare 500 ml
of 0.100 M HCl?

33. How much water must be added to 50.0 ml of 1.00 N NaOH

to produce 0.100 N NaOH? _____

New Terms

volumetric analysis   indicator
titration   end point
titrant   equivalence point
buret

Answers to Self-Test Questions

1. a, d   2. sugar, water   3. $KNO_3(s) \longrightarrow K^+(aq) + NO_3^-(aq)$
4. hydronium ion
5. (a) $NH_3 + H_2O \rightleftharpoons NH_4^+ + OH^-$

   (b) $HCN + H_2O \rightleftharpoons H_3O^+ + CN^-$

6. (a) S, rules 1 and 3   (b) S, rule 5   (c) I, rule 4   (d) I, rule 8   (e) I, rule 6   (f) I, rule 5   (g) I, rule 7   (h) S, rule 2

7. (a) $Na^+ + I^- + Ag^+ + NO_3^- \longrightarrow AgI + Na^+ + NO_3^-$

   $Ag^+ + I^- \longrightarrow AgI$

   (b) $Pb^{2+} + 2NO_3^- + Ba^{2+} + 2OH^- \longrightarrow Pb(OH)_2 + Ba^{2+} + 2NO_3^-$

   $Pb^{2+} + 2OH^- \longrightarrow Pb(OH)_2$

   (c) $AgCl + Na^+ + Br^- \longrightarrow AgBr + Na^+ + Cl^-$
   $AgCl + Br^- \longrightarrow AgBr + Cl^-$

   (d) $Zn^{2+} + 2Cl^- + 2Na^+ + CO_3^{2-} \longrightarrow 2Na^+ + 2Cl^- + ZnCO_3$

   $Zn^{2+} + CO_3^{2-} \longrightarrow ZnCO_3$

   (e) $Ba^{2+} + 2Cl^- + 2Na^+ + 2Br^- \longrightarrow Ba^{2+} + 2Br^- + 2Na^+ + 2Cl^-$
   (no net reaction)

8. (a) $CaCO_3 + 2H^+ + 2Cl^- \longrightarrow H_2O + CO_2 + Ca^{2+} + 2Cl^-$

   $CaCO_3 + 2H^+ \longrightarrow H_2O + CO_2 + Ca^{2+}$

   (b) $2NH_4^+ + SO_4^{2-} + 2Na^+ + 2OH^- \longrightarrow 2NH_3 + 2H_2O + 2Na^+ + SO_4^{2-}$

   $2NH_4^+ + 2OH^- \longrightarrow 2NH_3 + 2H_2O$; dividing through by 2
   gives $NH_4^+ + OH^- \longrightarrow NH_3 + H_2O$

(c) $2Na^+ + C_2O_4^{2-} + 2H^+ + 2Cl^- \longrightarrow 2Na^+ + 2Cl^- + H_2C_2O_4$

$C_2O_4^{2-} + 2H^+ \longrightarrow H_2C_2O_4$

(d) $BaCO_3 + 2H^+ + SO_4^{2-} \longrightarrow BaSO_4 + H_2O + CO_2$

(net is same as above)

(e) $2K^+ + SO_3^{2-} + 2H^+ + SO_4^{2-} \longrightarrow 2K^+ + SO_4^{2-} + H_2O + SO_2$

$SO_3^{2-} + 2H^+ \longrightarrow SO_2 + H_2O$

9. (a) $NiCl_2 + Na_2CO_3 \longrightarrow 2NaCl + NiCO_3$

$Ni^{2+} + 2Cl^- + 2Na^+ + CO_3^{2-} \longrightarrow 2Na^+ + 2Cl^- + NiCO_3(s)$

$Ni^{2+} + CO_3^{2-} \longrightarrow NiCO_3(s)$

(b) $2NaCl + CaBr_2 \longrightarrow 2NaBr + CaCl_2$

$2Na^+ + 2Cl^- + Ca^{2+} + 2Br^- \longrightarrow 2Na^+ + 2Br^- + Ca^{2+} + 2Cl^-$

(no net reaction)

(c) $MgCl_2 + 2NaOH \longrightarrow Mg(OH)_2 + 2NaCl$

$Mg^{2+} + 2Cl^- + 2Na^+ + 2OH^- \longrightarrow Mg(OH)_2(s) + 2Na^+ + 2Cl^-$

$Mg^{2+} + 2OH^- \longrightarrow Mg(OH)_2(s)$

(d) $Na_2SO_4 + Mg(NO_3)_2 \longrightarrow 2NaNO_3 + MgSO_4$

$2Na^+ + SO_4^{2-} + Mg^{2+} + 2NO_3^- \longrightarrow 2Na^+ + 2NO_3^- + Mg^{2+} + SO_4^{2-}$

(no net reaction)

(e) $Na_2S + 2HCl \longrightarrow 2NaCl + H_2S$

$2Na^+ + S^{2-} + 2H^+ + 2Cl^- \longrightarrow 2Na^+ + 2Cl^- + H_2S(g)$

$S^{2-} + 2H^+ \longrightarrow H_2S(g)$

10. (a) $Ca(OH)_2 + 2HCl \longrightarrow CaCl_2 + 2H_2O$

(b) $2NH_3 + H_2SO_4 \longrightarrow (NH_4)_2SO_4$

(c) $2Al(OH)_3 + 3H_2SO_4 \longrightarrow Al_2(SO_4)_3 + 6H_2O$

(d) $NaOH + HNO_3 \longrightarrow NaNO_3 + H_2O$

(e) $MgO + H_2SO_4 \longrightarrow MgSO_4 + H_2O$

11. (a) CaC_2O_4, $Ca(HC_2O_4)_2$ (b) $CaCO_3$, $Ca(HCO_3)_2$

(c) $Ca_3(AsO_4)_2$, $CaHAsO_4$, $Ca(H_2AsO_4)_2$ (d) $CaBr_2$

12. (a) acidic (SO_3 is nonmetal oxide)   (b) basic (BaO is metal oxide)   (c) acidic   (d) acidic

13. One possible set of reactants for each is given here. However, there is more than one way to "skin a cat." If you have chosen different reactants, ask your teacher if they are satisfactory.

(a) $FeCl_3(aq) + 3AgNO_3(aq) \longrightarrow Fe(NO_3)_3(aq) + 3AgCl(s)$

(b) $AgNO_3(aq) + KBr(aq) \longrightarrow AgBr(s) + KNO_3(aq)$

    in general: $Ag^+ + Br^- \longrightarrow AgBr(s)$

(c) $Ba(OH)_2(aq) + MgBr_2(aq) \longrightarrow BaBr_2(aq) + Mg(OH)_2(s)$

(d) $Ba(OH)_2(aq) + Na_2SO_4(aq) \longrightarrow BaSO_4(s) + 2NaOH(aq)$

14. (a) $CuCl_2(aq) + 2NaOH(aq) \longrightarrow Cu(OH)_2(s) + 2NaCl(aq)$

    $Cu(OH)_2(s) + 2H_2SO_4 \longrightarrow Cu(HSO_4)_2(aq) + 2H_2O$

(b) $NiCl_2(aq) + 2NaOH(aq) \longrightarrow 2NaCl(aq) + Ni(OH)_2(s)$

    $Ni(OH)_2(s) + H_2SO_4(aq) \longrightarrow NiSO_4(aq) + 2H_2O$

(c) $MgCl_2(aq) + 2NaOH(aq) \longrightarrow Mg(OH)_2(s) + 2NaCl$

    $Mg(OH)_2(s) + 2HNO_3(aq) \longrightarrow Mg(NO_3)_2(aq) + 2H_2O$

(d) $CaO(s) + 2HNO_3(aq) \longrightarrow Ca(NO_3)_2(aq) + H_2O$

15. (a) $FeCO_3(aq) + 2HCl(aq) \longrightarrow FeCl_2(aq) + H_2O + CO_2(aq)$

(b) $CoCO_3(aq) + 2HNO_3(aq) \longrightarrow Co(NO_3)_2(aq) + H_2O + CO_2(g)$

(c) $NH_4NO_2(aq) + NaOH(aq) \longrightarrow NaNO_2(aq) + H_2O + NH_3(g)$

(d) $ZnS(s) + 2HCl(aq) \longrightarrow ZnCl_2(aq) + H_2S(g)$

(e) $CaCl_2(aq) + 2NaOH(aq) \longrightarrow CaCO_3(s) + 2NaCl(aq)$

    $CaCO_3(s) + 2HNO_3(aq) \longrightarrow Ca(NO_3)_2(aq) + H_2O + CO_2(g)$

16. (a) $PH_3 + 4N_2O \longrightarrow H_3PO_4 + 4N_2$

(b) $NaIO_3 + 3Na_2SO_3 \longrightarrow 3Na_2SO_4 + NaI$

(c) $PbO_2 + 4HCl \longrightarrow PbCl_2 + 2H_2O + Cl_2$

17. (a) $2H^+ + 2HNO_2 + 2I^- \longrightarrow 2NO + I_2 + 2H_2O$

(b) $ClO_3^- + 3H_2S \longrightarrow Cl^- + 3S + 3H_2O$

(c) $8H_2O + 5S_2O_8^{2-} + 2P \longrightarrow 10SO_4^{2-} + 2H_3PO_4 + 10H^+$

18. (a) $H_2O + 2CrO_4^{2-} + 3SO_3^{2-} \longrightarrow 2CrO_2^- + 3SO_4^{2-} + 2OH^-$

(b) $OH^- + HO_2^- + 2ClO_2 \longrightarrow 2ClO_2^- + O_2 + H_2O$

(c) $H_2O + 2MnO_4^- + 3NO_2^- \longrightarrow 2MnO_2 + 3NO_3^- + 2\,OH^-$

(d) $ClO^- + 2NH_3 \longrightarrow N_2H_4 + Cl^- + H_2O$

19.(a) 0.760 M   (b) 0.480 M   (c) 1.20 M   (d) 0.471 M

20. 0.0750 mol $KClO_3$ 21. 0.0800 mol urea   22. 28.8 g of urea

23. 90.0 ml   24. 0.0720 mol AgCl (HCl is in excess)

25.(a) 140 g $NaBiO_3$ (b) 33.0 g $NaIO_3$ (c) 49.0 g H_2SO_4

(d) 98.1 g H_2SO_4 (e) 158 g $Na_2S_2O_3$

26. 16.4 g $Na_2S_2O_4$ (eq.wt. $Na_2S_2O_4$ = 87.06 g; $CuSO_4$ = 79.8 g)

27. 1.94 g K_2CrO_4 28. 42.9 ml   29. 0.18 N   30. 55.3% $CaCl_2$

31. 62.3% Fe   32. 16.7 ml of 3.00 M HCl

33. 450 ml water added (total volume = 500 ml)

7 GASES

In this chapter we examine the physical and chemical behavior of gases. This includes the way the properties of a gas depend on pressure, volume, and temperature. We will see that it is a relatively simple matter to measure the molecular weights of gaseous substances. The study of gases has also led to knowledge about the microscopic behavior of gases, including information about molecular size and the kinds of attractive forces that exist between gaseous atoms and molecules.

7.1 VOLUME AND PRESSURE

Objectives

To learn how pressure and volume are defined. You should learn how the pressure of a gas is measured and the units in which pressure is expressed.

Review

Gases expand to fill whatever container they are placed in. The volume of a gas is therefore the volume of its container.

Pressure is force per unit area. In the English system pressure can be expressed in pounds per square inch (psi). The pressure of a gas is usually given in terms of the height of a mercury column that exerts the same pressure as the gas. Remember that a mercury column 1 mm high exerts a pressure of 1 torr. At sea level, the pressure exerted by the atmosphere fluctuates about 760 torr. One standard atmosphere (1 atm) is defined as precisely 760 torr. The SI unit of pressure is the pascal (Pa). A standard atmosphere = 101,325 Pa (although we won't be using pascals in our calculations). You should be able to convert pressures from torr to atm, and vice versa using the relationship, 1 atm = 760 torr.

Gas pressures are normally measured using manometers. In the text you see that for an open-end manometer the equation used to calculate the gas pressure depends on whether the gas pressure is greater than or less than the atmospheric pressure. Don't try to memorize these equations - it's too easy to get them confused. It is much better if you can learn how to analyze the manometer. Remember that the lower of the two liquid levels is always chosen as the reference level. At this level the pressures on both the left and right sides are the same. Figure out what they are, equate them, and then solve for the pressure of the gas.

Sometimes a liquid other than mercury is used in a manometer. To express the pressure in torr it is necessary to convert the difference in heights of the liquid columns to the difference in heights that would have been found had mercury been in the manometer. This is obtained by multiplying the difference in heights of the liquid by a ratio of the densities of the liquid and mercury. Mercury is the most dense liquid ever used in a manometer. Therefore, the height of the mercury column will always be less than the height of the liquid column. Always set up the ratio of densities so that the calculated equivalent column of mercury is less than that of the liquid. Review Example 7.1 in the text and Example 7.1 below.

Example 7.1

A liquid having a density of 1.22 g/ml is used in a manometer. In measuring the pressure of a gas, a difference of 15.6 cm was observed between the liquid levels in the two arms. What is the mercury equivalent, in torr, of this column of liquid. The density of mercury is 13.6 g/ml.

Solution

The mercury equivalent of the liquid column is obtained by

multiplying the height of the liquid column (15.6 cm) by a ratio of densities.

$$P_{Hg} = P_{liquid} \text{ x (ratio of densities)}$$

Since P_{Hg} will be less than P_{liquid}, the ratio of densities must have some value smaller than 1, so that when P_{liquid} is multiplied by this ratio a smaller value is obtained. A value less than 1 is only obtained as 1.22/13.6. Therefore,

$$P_{Hg} = 15.6 \text{ cm } x \left(\frac{1.22}{13.6}\right) = 1.40 \text{ cm Hg}$$

Since 1 cm = 10 mm and 1 torr = 1 mm Hg,

$$P_{Hg} = 14.0 \text{ mm Hg} = 14.0 \text{ torr}$$

Self-Test

1. If a gas exerts a pressure of 1.35 atm, what is its pressure in torr?

   _____

2. If a gas exerts a pressure of 630 torr, what is its pressure in atm?

   _____

3. A gas in a container is attached to an open-end manometer filled with mercury. The mercury level in the arm connected to the container is 18.5 mm below the level in the open arm. The atmospheric pressure is 755.3 torr. Calculate the pressure of the gas. (It will help if you sketch a picture of the apparatus.)

   _____

4. A gas contained in a vessel exerts a pressure of 1.14 atm. The vessel is connected to an open-end manometer filled with mercury. The atmospheric pressure is 766 torr. What will be the difference in heights (measured in cm) between the levels of mercury in the two arms?

   _____

5. Two gases in the cylinders shown in Figure 7.1 (next page) exert pressures of 785 torr and 790 torr. If the manometer is filled with water (d = 1.00 g/ml), what will be the difference in heights between the liquid levels?

   _____

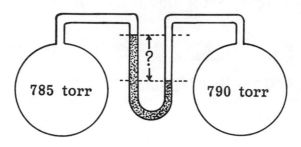

Figure 7.1

6. Calculate the pressure, in atm, exerted on a skin diver at a depth of 12 fathoms on a day when the atmospheric pressure is 752 torr (a barometric pressure of 29.6 inches of mercury). The density of sea water is 1.02 g/ml; 1 fathom equals 6 feet.

_____

New Terms

barometer   torr
standard atmosphere   manometer
pascal

7.2   BOYLE'S LAW

Objectives

To learn how the volume of a gas varies with the pressure exerted on it when the amount of gas and its temperature are held constant.

Review

Remember that as the pressure on a gas increases, the gas is squeezed into a smaller volume. Boyle's law says that at constant temperature the product, PV, for a fixed quantity of gas is constant. For real gases this is not quite true, although near atmospheric pressure and room temperature most gases follow Boyle's law quite well. An ideal gas is a hypothetical gas that would obey Boyle's law perfectly under all conditions.

You will be expected to be able to perform calculations dealing with Boyle's law. There are two ways to approach these problems:

(1) You can memorize the equation, $P_1V_1 = P_2V_2$. This equation can be used, for example, to calculate the final pressure of a gas if you know its initial pressure and volume, and its final volume.

(2) You can use the idea that the final volume of a gas is equal to its initial volume multiplied by a ratio of pressures. For example, an increase in pressure causes the final volume to be smaller than the initial volume. The ratio of pressures must therefore be smaller than one so that when the initial volume is multiplied by the ratio the result is a smaller volume. Since this reasoning approach requires some practice, another example is presented below.

Example 7.2

A gas, initially occupying 2.00 liters at 780 torr, is placed in a container (at the same temperature) in which its pressure is 740 torr. What is the volume of the container?

Solution

Remember to set up a table of initial and final conditions to avoid confusion.

	initial	final
P	780 torr	740 torr
V	2.00 liters	?

Next, reason through the problem. The pressure on the gas has decreased ($780 \rightarrow 740$). Therefore, the gas must have expanded. This means that the final volume must be larger than 2.00 liters. Which ratio of pressures do we choose to multiply the initial volume by?

$$\frac{780 \text{ torr}}{740 \text{ torr}} \quad \text{or} \quad \frac{740 \text{ torr}}{780 \text{ torr}}$$

Obviously (780/740) is the correct choice since it is the only one that will make the final volume larger. The answer is then obtained as

$$V_f = 2.00 \text{ liters} \times \left(\frac{780 \text{ torr}}{740 \text{ torr}}\right) = 2.11 \text{ liters}$$

As a final note, remember that it is absolutely necessary to have the units of the two quantities (P or V) in a ratio the same. The units in the ratio must cancel.

Self-Test

7. A gas at 745 torr occupies 250 ml. What volume will it occupy at 300 torr if the temperature remains the same?

_____

8. A gas occupies a volume of 180 ml at a pressure of 450 torr. If the gas is transferred to a 2.00-liter container at the same temperature, what will be the new pressure?

_____

9. Gasoline vapor mixed with air at atmospheric pressure (1 atm) is drawn into a 610-ml cylinder in an auto engine. Before the mixture is exploded the piston compresses the gas to a volume of 73.1 ml. What is the pressure of the gas when the mixture is ignited? (Assume no temperature change.)

_____

New Terms

Boyle's law
ideal gas

7.3   CHARLES' LAW

Objectives

To learn how the volume of a gas is related to its temperature.

Review

Remember that absolute zero occurs at -273°C. Temperatures used in gas law calculations are always expressed in kelvins.

The kelvin temperature is obtained by adding 273 to the Celsius temperature.

$$K = °C + 273$$

Calculations involving Charles' law, like those with Boyle's law, can also be approached in two ways:

(1) You can use the equation,

$$\frac{V_1}{T_1} = \frac{V_2}{T_2}$$

(2) You can use the fact that as the temperature of a gas increases, at constant pressure, its volume increases. Stated more simply, the gas expands as it gets hot.

As before we have a situation where the final volume is equal to the initial volume multiplied by a ratio of absolute temperatures, or the final absolute temperature is equal to the initial absolute temperature multiplied by a volume ratio.

Example 7.3

A gas occupies 250 ml at 25°C. If the pressure remains constant, at what temperature (in °C) will the gas occupy 300 ml?

Solution

First we collect the data in a table and convert the temperature to kelvins.

	initial	final
V	250 ml	300 ml
T	25 + 273 = 298 K	?

We can compute the final temperature T_f as follows:

$$T_f = 298 \text{ K} \times \text{(ratio of volumes)}$$

Since the volume is increasing, the temperature must be rising because gases expand as they are heated. This means that $T_f > T_i$, so the initial temperature must be multiplied by a volume ratio that is larger than one.

$$T_f = 298 \text{ K } \times \left(\frac{300 \text{ ml}}{250 \text{ ml}}\right)$$

$$= 358 \text{ K}$$

Converting to Celsius (by subtracting 273), the final temperature is 85°C.

Self-Test

10. A gas occupies a volume of 650 ml at 30°C. What volume will it occupy at 100°C, assuming that its pressure remains constant?

11. A gas occupies a volume of 26.5 ml at 300°C. While being held at constant pressure, its volume decreases to 15.7 ml. What has the temperature of the gas become (in °C)?

New Terms

Charles' law   absolute temperature
absolute zero

7.4   GAY-LUSSAC'S LAW

Objectives

To learn how the pressure of a gas is related to its temperature.

Review

Warming a confined gas increases its pressure. Thus

$$P \propto T$$

or

$$\frac{P_1}{T_1} = \frac{P_2}{T_2}$$

As with the other gas laws, we can approach problems involving

Gay-Lussac's law by the reasoning method. This is shown in Example 7.4 in the text.

Self-Test

12. A gas in a container of fixed volume exerts a pressure of 855 torr at 350°C. When the temperature of the gas is changed the pressure rises to 980 torr. What is the new temperature of the gas (in °C)?

_____

13. A sample of methane (natural gas) exerts a pressure of 350 torr at a temperature of 18°C. What pressure will this sample exert if its volume is held constant and its temperature is raised to 45°C?

_____

New Terms

Gay-Lussac's law

7.5   THE COMBINED GAS LAW

Objectives

> To combine the preceding three gas laws into one expression. You should learn the meaning of standard temperature and pressure.

Review

> Reference conditions for gases have been chosen to be 0°C (standard temperature) and 760 torr (standard pressure). STP is the abbreviation used to specify standard temperature and pressure.

> As with the other gas laws, calculations with the combined gas law can be solved using the reasoning approach; for example,

$$V_f = V_i \text{(ratio of pressures)(ratio of temperatures)}$$

You can also use the equation,

$$\frac{P_1 V_1}{T_1} = \frac{P_2 V_2}{T_2}$$

Review Examples 7.5 and 7.6 in the text.

Self-Test

14. A gas at 25°C and 740 torr occupies 840 ml. What volume will it occupy at 14°C and 650 torr?

    _____

15. A sealab having an open hatch at the bottom is submerged to a depth at which the pressure is 8.5 atm and the temperature is 13°C. At the surface (P = 1 atm) its volume is 5000 ft^3 and the temperature is a balmy 30°C. If no air is pumped into the sealab, how much usable air space will remain when it is submerged?

    _____

16. A sample of air at 35°C has a volume of 365 ml at a pressure of 850 torr. If it is transferred to a 280-ml container at a temperature of 50°C, what will be its pressure?

    _____

17. If a sample of a gas occupies 265 ml at 920 torr and 37°C, what volume will it occupy at STP?

    _____

New Terms

combined gas law   standard pressure
standard temperature   STP

7.6 DALTON'S LAW OF PARTIAL PRESSURES

Objectives

To observe how gases behave in mixtures.

Review

Dalton's law is very simple. Each gas in a mixture behaves independently of any other gases present and exerts a pressure (called its partial pressure) that is the same as it would exert if it were alone in the container. The total pressure of the mixture is simply the sum of the pressures of each of the gases.

A practical application of Dalton's law is the calculation of the pressure of a gas collected over a liquid such as water. Review Examples 7.7 and 7.8 in the text before attempting the Self-Test below.

Self-Test

18. 300 ml of argon at a pressure of 420 torr and 300 ml of helium at 240 torr are placed into the same 300-ml container. The temperatures of the separate gases and the mixture are the same. What is the partial pressure of each gas in the mixture and the total pressure of the mixture?

19. 200 ml of N_2 at 30°C and 750 torr is mixed with some O_2 and transferred to a 500-ml container at 30°C. The total pressure of the mixture is found to be 680 torr. What is the partial pressure of each gas in the mixture?

20. 400 ml of oxygen was collected over water at 25°C at a total pressure of 765 torr. What is the partial pressure of the trapped oxygen?

New Terms

partial pressure
Dalton's law of partial pressure
vapor pressure

7.7   CHEMICAL REACTIONS BETWEEN GASES

Objectives

To learn the quantitative relationships between volumes of gases involved in chemical reactions. You should learn the volume occupied by one mole of gas at STP.

Review

Gay-Lussac's law of combining volumes states that at constant temperature and pressure the volumes of gases consumed and/or produced in a chemical reaction are related to one another as ratios of small whole numbers. In fact, under conditions of constant T and P, the volume ratios are the same as the ratios of the coefficients in the balanced chemical equation.

Also of historical importance is Avogadro's principle - equal volumes of gas at the same T and P contain equal numbers of molecules. This can also be expressed as <u>the volume of a gas at constant temperature and pressure is proportional to the number of moles of gas present</u>. For an ideal gas, the volume occupied by one mole at STP (the molar volume of the gas) is 22.4 liters.

Self-Test

21. What volume of O_2 is required to completely react with 150 ml of C_2H_6 (both volumes measured at the same T and P) according to the equation
$$2C_2H_6(g) + 7O_2(g) \longrightarrow 4CO_2(g) + 6H_2O(g) ?$$

_____

22. What volume of air (composed of 20% O_2 by volume) at 25°C and 1.00 atm is required to react with 500 ml of C_2H_6 (measured at 30°C and 850 torr)? (This problem requires several steps - think about how to solve the problem before working with the numbers.)

_____

23. What volume (in liters) would 24.6 g of C_2H_6 occupy at STP?

_____

24. How many liters of oxygen, measured at STP, are needed to oxidize 14.6 g of iron according to the equation
$$4Fe + 3O_2 \longrightarrow 2Fe_2O_3 ?$$

_____

New Terms

Gay-Lussac's law of combining volumes
Avogadro's principle

7.8   THE IDEAL GAS LAW

Objectives

To obtain an equation that encompasses all of the gas laws that we have examined in previous sections.

Review

You should learn the ideal gas law,

$$PV = nRT$$

The gas constant, R, can have different numerical values depending on the units used to express pressure and volume. The value used most frequently in the text is 0.0821 liter atm/mol K. If you learn this value, it's important that you also learn the units that go with it. If you use this value in the ideal gas law, remember that the pressure must be expressed in atm and the volume in liters; otherwise incorrect numerical answers will result regardless of how good you are at arithmetic.

A very important application of the ideal gas law is in the determination of molecular weights of gaseous substances. Review Examples 7.13 and 7.14 in the text as well as the Example below.

Example 7.4

A sample of a gas was found to have a density of 1.64 g/liter at 30°C and 0.930 atm. What is the molecular weight of the gas?

Solution

When you are asked to compute a molecular weight, you actually are being asked to calculate the number of grams per mole. The necessary data for this are P, V, T and a weight of gas (in grams). From the P, V, T data you can calculate the number of moles of gas from the ideal gas law. The molecular weight is then obtained simply by taking the ratio of grams of gas/moles of gas.

The density in this question gives the weight of one liter of gas. Therefore, we have: P = 0.930 atm, V = 1.00 liter, T = 273 + 30 = 303 K, mass = 1.64 g. Solving the ideal gas law for n,

$$n = \frac{PV}{RT} = \frac{(0.930 \text{ atm})(1.00 \text{ liter})}{(0.0821 \text{ liter atm/mol K})(303 \text{ K})}$$

$$n = 0.0374 \text{ mol}$$

Then,

$$M.W. = \frac{1.64 \text{ g}}{0.0374 \text{ mol}} = 43.9 \text{ g/mol}$$

___

Self-Test

25. Calculate the numerical value of R having the units,

    (a) ml atm/mol K   _____

    (b) liter torr/mol K   _____

26. How many moles of gas are present in 3.00 liters at 800 torr and 40°C?

    _____

27. Calculate the pressure in atm exerted by 0.10 mol of argon at -20°C in a 10-liter container.

    _____

28. 0.625 g of an unknown gas occupies 500 ml at STP. What is its molecular weight?

    _____

29. What volume would 28 g of O_2 occupy at 800 torr and 27°C?

    _____

30. Calculate the density of N_2 at STP.   _____

31. Butane, from a cigarette lighter, has a density of 2.30 g/liter at 22°C and 730 torr. What is the molecular weight of butane?

    If its empirical formula is C_2H_5, what is its molecular formula?

    _____

New Terms

ideal gas law   universal gas constant
equation of state for an ideal gas

7.9   GRAHAM'S LAW OF EFFUSION

Objectives

To learn how the rates of effusion (and diffusion) of gases are related to their molecular weights.

Review

Diffusion and effusion are similar, although not identical, processes. Both refer to the rates at which gas molecules move from one place to another. Graham's law is summarized in the equation,

$$\frac{\text{rate of effusion of gas (A)}}{\text{rate of effusion of gas (B)}} = \sqrt{\frac{M_B}{M_A}}$$

where M_A and M_B are the molecular weights of A and B, respectively.

Self-Test

32.   Which of the following molecules diffuses faster?

(a) H_2O or H_2S _____

(b) NH_3 or H_2O _____

(c) CO_2 or NO_2 _____

33.   Calculate the ratio of the rates of effusion, R_{NO}/R_{NO_2}.

_____

34.   How many times faster does $^{235}UF_6$ diffuse than $^{238}UF_6$?

_____

35.   What would the molecular weight of a gas be if it diffuses only one sixth as fast as H_2?

_____

New Terms

diffusion   Graham's law
effusion

7.10   KINETIC MOLECULAR THEORY AND THE GAS LAWS

Objectives

To learn and understand the postulates of the theory that
was developed to explain the gas laws which we discussed
in earlier sections. You should learn how the distribution
of molecular speeds and molecular kinetic energies are related to the absolute temperature.

Review

The kinetic molecular theory was developed to explain why
gases behave the way they do. The theory is described by a set
of postulates presented in detail in the text. In summary, the
kinetic molecular theory (or simply the kinetic theory) views a
gas as being composed of very tiny molecules, having negligible
volume themselves, separated by very large distances from one
another. The molecules are in rapid random motion, colliding
with the walls of the container and with each other. The pressure exerted by a gas results from collisions of the molecules
with the walls. It is further postulated that the molecules do not
attract each other and that there is a distribution of molecular
speeds, ranging from very slow molecules to extremely fast ones.
Associated with the distribution of molecular speeds there is a
corresponding distribution of kinetic energies. To account for
Graham's law it is necessary to postulate that the average kinetic
energy of a gas depends only on the absolute temperature. This
is perhaps the most important aspect of the kinetic theory because
it applies to any collection of molecules. At a given temperature
the average kinetic energy is the same for any collection of molecules, regardless of their chemical makeup or whether they are
in a gas, a liquid, or a solid. Study Figure 7.12. In particular,
note the following:

(1)   At any temperature the fraction of molecules having zero
      kinetic energy is zero.

(2) The fraction of molecules having very large kinetic energies gradually approaches zero at high kinetic energies.

(3) The maximum on the curve at a given temperature represents the kinetic energy possessed by the largest fraction of molecules. This is termed the most probable kinetic energy since it is the one most likely to be found if the molecules were sampled at random.

(4) At successively higher temperatures the height of the maximum decreases. The total area under the curve represents the sum of all of the fractions, and must equal 1.00. Since the curve gets higher at large kinetic energies, it must get lower elsewhere so that the area remains constant.

(5) At any given temperature the average kinetic energy occurs at a slightly higher kinetic energy than the most probable kinetic energy. This is a consequence of the unsymmetrical shape of the distribution curve.

(6) The average kinetic energy increases as the temperature increases.

Self-Test

36. (Multiple choice). How does kinetic theory account for Boyle's law?
     (a) The average kinetic energy depends only on temperature.
     (b) A gas is mostly empty space.
     (c) There are no attractive forces between molecules.
     (d) The molecules are in rapid random motion.
     _____

37. (Multiple choice). How does kinetic theory account for Charles' law?
     (a) There is a distribution of molecular speeds between gas molecules.
     (b) The molecules have negligibly small volumes.
     (c) Molecules collide with each other.
     (d) On the average, molecules move faster at higher temperature.
     _____

38. After studying this section, sketch the kinetic energy distribution for a gas at two different temperatures on the axes which follow. Indicate the average and most probable kinetic

energies at both temperatures.

New Terms

kinetic molecular theory
mole fraction

7.11   REAL GASES

Objectives

To see how the postulates of the kinetic theory must be modified to account for the properties of real gases, which do not obey the ideal gas law exactly.

Review

There are two defects in the kinetic theory presented in Section 7.10.

(1)   Molecules do have attractive forces between them.   An example is the dipole-dipole attractions between polar molecules.

(2)   Molecules themselves do have a finite volume that is not negligible compared to the total volume when the molecules are squeezed close together.

The van der Waals equation,

$$\left(P + \frac{n^2 a}{V^2}\right)\left(V - nb\right) = nRT$$

attempts to apply corrections to the pressure and volume of a gas in the ideal gas law. The actual pressure and volume are modified to give a pressure and volume that the gas would have if it were an ideal gas (i.e., if there were no attractive forces and if the gas molecules had zero volume). In the equation, the constant a is proportional to the strengths of the attractive forces and b is proportional to the size of the molecules.

Self-Test

39. Use the data in Table 7.3 to answer the following:

   (a) Which gas has the larger molecules, CH_4 or H_2O?_____

   (b) Which gas has the greater attractive forces between molecules, NH_3 or H_2O?

   _____

   (c) Which gas has the larger molecules, H_2O or C_2H_5OH?

   _____

   (d) Which gas has the greater attractive forces, O_2 or CH_4?

   _____

New Terms

excluded volume
van der Waals equation of state

Answers to Self-Test Questions

1. 1026 torr   2. 0.829 atm   3. 773.8 torr   4. 100 mm   5. 68 mm
6. 3.16 atm (2.17 atm from sea water, 0.99 atm from atmosphere)
7. 621 ml   8. 40.5 torr   9. 8.34 atm   10. 800 ml   11. 66°C
12. 441°C   13. 382 torr   14. 921 ml   15. 555 ft^3 (This question illustrates why underwater laboratories open to the sea are pressurized - to keep the water out!)   16. 1.16 x 10^3 torr
17. 283 ml   18. P_{Ar} = 420 torr, P_{He} = 240 torr, P_T = 660 torr

19. p_{N_2} = 300 torr, p_{O_2} = (680 - 300) = 380 torr

20. 741 torr   21. 525 ml O_2 22. 9.62 liters   23. 18.4 liters
24. 4.39 liters   25.(a) 82.1 ml atm/mol K (b) 62.4 liter torr/mol
K   26. 0.123 mol   27. 0.21 atm   28. 28.0 g/mol
29. 20 liters   30. 1.25 g/liter   31. 58.0 g/mol, C_4H_{10}
32.(a) H_2O (b) NH_3 (c) CO_2: The lighter (lower molecular

weight) molecule diffuses faster.   33. 1.24   34. 1.004
35. 72 amu   36. b   37. d   38. See Figure 7.12.
39. (a) CH_4 (b) H_2O (c) C_2H_5OH (d) CH_4

8 STATES OF MATTER AND INTERMOLECULAR FORCES

This chapter focuses on the physical properties of liquids and solids, and how these properties are affected by the attractive forces between the particles (molecules or ions) of which they are composed. As you study this chapter you will learn the reasons for many of the properties of gases, liquids, and solids that we take for granted on a day-to-day basis. Perhaps you will see the world around you in a new light.

8.1 COMPARING THE PROPERTIES OF GASES, LIQUIDS, AND SOLIDS

Objectives

To learn why gases, liquids, and solids differ so greatly in their properties and to examine some specific properties of liquids and solids.

Review

The physical properties of all gases are nearly alike because the molecules are very far apart and the attractive forces between them are very weak. In a liquid or solid, the particles are very close together, with very little empty space. In addition, in a liquid or solid the attractive forces are relatively strong and depend on the kinds of particles. For this reason, different chemical substances behave differently in their liquid and solid states.

Compressibility and rates of diffusion are properties that are determined primarily by the tightness of packing in the various states.

Volume and shape depend on the strengths of the attractive forces. In a solid the attractions between molecules or ions prevent them from easily moving past each other.

Surface tension is the energy needed to increase the surface area of a liquid and depends strongly on the attractions between the molecules. Liquids of low surface tension are able to wet surfaces easily.

The rate of evaporation of a liquid or solid depends on the strengths of intermolecular attractions and increases with increasing temperature. It also depends on its surface area. Be sure to study Figures 8.6 and 8.7, with their accompanying explanations. Remember that evaporation is an endothermic process, and produces a cooling effect.

Self-Test

1. (Multiple choice). The fact that moist grains of sand stick together is attributed to liquid water's
   (a) small molecular size
   (b) surface tension
   (c) inability to be compressed
   (d) ability to change its shape   _____

2. (Multiple choice). Water evaporates faster at high temperatures than at low temperatures primarily because
   (a) increasing the temperature of a liquid causes it to expand
   (b) at high temperature the molecules are further apart
   (c) water wets a surface faster at high temperature
   (d) at the higher temperature more molecules have high kinetic energies

   _____

3. Alcohol evaporates faster than water when they are at the same temperature.
   (a) Which substance has the weaker attractive forces between its molecules?

   _____

   (b) Which would be expected to have the lower surface tension?

   _____

4. (Multiple choice). The reason molecules diffuse more slowly in liquids than in gases is
    (a) the molecules move more slowly in a liquid than in a gas
    (b) the strong attractive forces in a liquid hold the molecules in place
    (c) the molecules move slowly because the liquid cannot expand easily
    (d) the molecules are constantly colliding with others, thereby interfering with their movement

    _____

5. How does a surfactant increase the ability of water to wet a surface?

    _____

    _____

New Terms

compressibility
mean free path
surface tension
wetting

surfactant
evaporation
sublimation

8.2   INTERMOLECULAR ATTRACTIVE FORCES

Objectives

To learn about the kinds of intermolecular attractive forces and their relative strengths.

Review

Intermolecular attractions are the attractive forces between neighboring molecules. There are three principal types: dipole-dipole attractions, hydrogen bonds, and London forces.

Dipole-dipole attractions occur between polar molecules and are generally about 1% as strong as normal covalent bonds. Hydrogen bonding is an especially strong dipole-dipole attraction that occurs when hydrogen is bonded to a small, very electronegative element - principally fluorine, oxygen, and nitrogen. Their strengths are about 5% to 10% as strong as covalent bonds.

London forces (instantaneous dipole-induced dipole forces) occur between all particles, but they are especially important in non-polar substances where they are the only intermolecular forces present.

Self-Test

6. What kinds of attractive forces occur in the liquid state between molecules of

    (a) HBr _____

    (b) CO_2 _____

    (c) H_2O _____

    (d) CH_3Cl _____

    (e) CH_3OH _____

7. Why are London forces normally weak compared to dipole-dipole forces?

    _____

_____

New Terms

dipole-dipole attractions   instantaneous dipole
hydrogen bonds   induced dipole
London forces

8.3   HEAT OF VAPORIZATION

Objectives

To examine the energy changes that accompany the evaporation of a liquid and to learn what they tell us about the strengths of intermolecular attractions.

Review

The molar heat of vaporization is the energy required to convert one mole of liquid to one mole of vapor. Quantitatively, it is the difference between the heat content of the vapor and the heat content of the liquid.

$$\Delta H_{vap} = H_{vapor} - H_{liquid}$$

Remember that neither H_{vapor} nor H_{liquid} can actually be measured; it is only their difference that is observed.

The heat of vaporization is useful because it provides a direct measure of the strengths of the attractive forces that exist between the molecules in the liquid.

Variations in ΔH_{vap} show that among hydrocarbons the attractive forces (London forces) increase with chain length. London forces also increase with molecular size because large molecules are more polarizable than small ones. In this section we see that hydrogen bonding is important for HF, H_2O and NH_3. In general, hydrogen bonding is most significant in molecules having O—H or N—H bonds.

Self-Test

8. Using the data in Table 8.1, arrange the following compounds in order of increasing strengths of intermolecular attractive forces: C_2H_6, HCl, H_2S, HF, SiH_4, NH_3.

   _____

9. Without referring to Table 8.1, choose the compound in each of the following pairs with the stronger intermolecular attractive forces.
   (a) PH_3 or AsH_3 _____
   (b) SiH_4 or CH_4 _____
   (c) H_2O or H_2S _____

New Terms

molar heat of vaporization
polarizability

8.4   VAPOR PRESSURES OF LIQUIDS

Objectives

To understand why the vapor pressure of a liquid depends only on the temperature. You should understand the concept of dynamic equilibrium and you should learn how the effects of outside influences on an equilibrium can be predicted by application of Le Châtelier's principle. In particular, you should learn how temperature affects the vapor pressure of a liquid.

Review

A dynamic equilibrium exists when two opposing processes occur at the same speed.

The equilibrium vapor pressure of a liquid (usually just called its vapor pressure) is the pressure exerted by its vapor when the vapor is in dynamic equilibrium with the liquid. In this case molecules are evaporating from the liquid into a closed container at the same rate that molecules are returning to the liquid.

Various factors can influence the position of equilibrium - that is, the relative amounts of reactants and products in a chemical system or, in this case, the relative amounts of liquid and vapor. Le Châtelier's principle states that when a system at equilibrium is disturbed (so as to upset the equilibrium) the system readjusts in a way that minimizes, or counteracts, the stress placed upon it. If the pressure on the system is increased by a decrease in volume, the system will respond (if it can) in a way that tends to reduce the pressure. If heat is added to a system, the system responds by undergoing a change that absorbs heat. In each case the system changes in a way that tends to absorb the stress placed on it. You will encounter Le Châtelier's principle again in Chapters 13, 15 and 16. Learning to apply it now will make things easier for you later on.

The vapor pressure of a liquid increases with temperature. A graph of vapor pressure versus temperature gives a vapor pressure curve, illustrated in Figure 8.15 in the text. The vapor pressure curve ends at the critical temperature - the temperature above which a gas can no longer be condensed to liquid by the application of pressure. At the critical temperature a gas can be liquefied by application of the critical pressure.

Self-Test

10. Consider the process, vapor $\rightleftharpoons$ liquid + heat

(a)   A decrease in pressure will increase the amount of

_____

(b)   An increase in temperature will decrease the amount of

_____

11.   (Multiple choice).   The vapor pressure of a liquid increases
with increasing temperature primarily because as the temperature rises,
(a)   the molecules of the vapor move more rapidly
(b)   a greater fraction of molecules can escape the liquid
(c)   the attractive forces between the molecules in the vapor
decrease
(d)   the rate of return to the liquid increases   _____

12.   Refer to Table 8.1 to answer this question.   Should HBr or
HCl be expected to have the larger vapor pressure at
-75°C?

_____

13.   Ethylene glycol has a very low vapor pressure at room temperature.
(a)   Does ethylene glycol evaporate rapidly at room temperature?

_____

(b)   What does the information in this question tell you
about the strengths of the attractive forces in ethylene
glycol?
_____

_____

New Terms

dynamic equilibrium   vapor pressure curve
equilibrium vapor pressure   critical temperature
Le Châtelier's principle   critical pressure

8.5   BOILING POINT

Objectives

To define more precisely the term, boiling point, and to
understand why boiling point changes with pressure.   You

should also learn how boiling point provides a measure of the strengths of the intermolecular attractive forces in a liquid.

Review

The boiling point is the temperature at which the vapor pressure of the liquid equals the prevailing atmospheric pressure. The normal boiling point (standard boiling point) is the temperature at which the vapor pressure equals 760 torr.

The boiling point provides an indication of the strengths of the attractive forces between liquid molecules. If the attractive forces are high, the vapor pressure at a given temperature is low because only a small fraction of molecules can escape the liquid. These liquids must be heated to high temperatures to bring their vapor pressures up to atmospheric pressure. On the other hand, when weak attractive forces are present, a large fraction can escape and the vapor pressure is high. Liquids that have low vapor pressures at a given temperature have high boiling points, while those with high vapor pressures have low boiling points.

The abnormally high boiling points of NH_3, H_2O and HF provide evidence for hydrogen bonding in these substances.

Self-Test

14. Why does HF have a lower boiling point than H_2O even though it forms stronger hydrogen bonds?

_____

_____

15. What effect does an increase in pressure have on the boiling point of a liquid? _____

16. What is inside the bubbles in boiling water? _____

New Terms

boiling point
normal boiling point

8.6   FREEZING POINT

Objectives

>   To define freezing point in terms of dynamic equilibrium
>   and to consider the energy changes that take place upon
>   freezing and melting.

Review

>   At the freezing point of a liquid there is a dynamic equi-
librium between molecules in the solid and liquid. Molecules leave
the solid and enter the liquid at the same rate that molecules
leave the liquid and attach themselves to the solid. The energy
that must be removed from one mole of liquid to convert it to solid is called the molar heat of crystallization. This is equal in
magnitude, but opposite in sign, to the molar heat of fusion -
the energy needed to melt one mole of solid. Fusion means the
same as melting (That is how an <u>electrical fuse</u> works.). Remember that ΔH_{fus} is always much less than ΔH_{vap}.

Self-Test

17. Why doesn't the value of ΔH_{fus} give a direct measure of the
     strengths of the attractive forces in the solid?

     _____

     _____

     _____

18. What difference is there between the freezing point of a
     liquid and the melting point of a solid?

     _____

New Terms

freezing point   molar heat of crystallization
melting point   molar heat of fusion
fusion

8.7   CRYSTALLINE SOLIDS

Objectives

To learn what features identify crystalline solids and to learn how their structures are investigated using X-ray diffraction.

Review

A crystal normally has a very regular, symmetrical form that is the result of the very orderly pattern of particles within it.

The structures of crystals are studied by X-ray diffraction. When an X-ray beam is directed on a crystal the atoms composing the crystal scatter the beam in all directions. The X rays emerging from the crystal are only in phase in certain directions, however, and an intense X-ray beam is observed to come out of the crystal only at certain angles with respect to the incoming beam. The key point in this section is that the distance of separation between planes of atoms in a crystal is related to the angle at which an X-ray beam is observed to be reflected.

The Bragg equation relates the angle of reflection (θ), the distance between planes of atoms (d) and the wavelength of the X rays (λ). Using X rays of known wavelength the distances between atoms in the crystal can be calculated. Bragg's equation is

$$n\lambda = 2d \sin \theta$$

where n is an integer (n = 1 or 2 or 3, etc.)

Self-Test

19. X rays of wavelength 154 pm are reflected from layers of atoms in a crystal of potassium chloride at an angle of 14.1°. What is the distance between the layers? (Assume n = 1)

_____

New Terms

crystalline solid   Bragg equation
X-ray diffraction

8.8   LATTICES

Objectives

To find a way of describing a repeating pattern, such as the structure of a crystal, in terms of a lattice.

Review

One of the main themes throughout this section is that it is possible to describe the structures of millions of different crystals in terms of a very small set of only 14 three-dimensional lattices.

A lattice may be a one-, two- or three-dimensional geometric array of points. To describe a crystal, of course, a three-dimensional lattice is required. The properties of the entire lattice are embodied in the unit cell. The entire lattice can be constructed by moving the unit cell along its edges by distances equal to the lengths of the edges. The unit cell is characterized by the lengths of its edges, a, b and c, and the angles opposite them, α, β and γ, respectively (see Figure 8.25 in the text).

Three common types of lattices are described by the cubic unit cells.

(1)   simple cubic (primitive cubic): lattice points only at the corners.

(2)   body-centered cubic: lattice points at the corners plus one in the center of the cell.

(3)   face-centered cubic: lattice points at the corners plus one in the center of each face.

Remember that only 1/8 of a corner point lies within a given unit cell, 1/4 of a point along an edge lies in a given cell, 1/2 of a face-centered point lies within a given cell, and any point within the cell contributes one entire point to the cell.

Measurement of unit cell dimensions allows the calculation of the sizes of atoms and ions. The procedure is essentially a problem in applied geometry, as shown on Page 270.

Self-Test

20. Consider the unit cell drawn below. How many lattice points
     are within it?

     _____

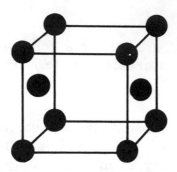

21. Cesium chloride, CsCl, forms crystals having a simple cubic
     lattice. If cesium ions are located at the corners of the unit
     cells, where must the choride ions be located?

     _____

22. Iron crystallizes in a body-centered cubic lattice, with atoms
     touching along the body diagonal of the cube - the line
     running from one corner through the center of the cube to
     the opposite corner. The atomic radius of iron is 126 pm.
     What is the length of the edge of the unit cell expressed in
     picometers?

     _____

23. Potassium chloride crystallizes with a face-centered cubic
     structure. One face of a unit cell is shown on p 153. The
     unit cell edge is 628 pm long and the radius of the Cl^- ion
     is 181 pm.

     (a) What is the ionic radius of K^+ in picometers?_____

     (b) What is the radius in angstroms?   _____

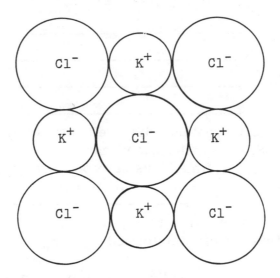

New Terms

lattice
crystal lattice
unit cell
simple cubic unit cell

primitive lattice
body-centered cubic unit cell
face-centered cubic unit cell

8.9   TYPES OF CRYSTALS

Objectives

To understand how the properties of a crystalline substance depend on the kinds of species that occupy sites
in the lattice and on the nature of the attractive forces
between them.

Review

The key points of this section are summarized in Table 8.5
in the text. Study this table well before answering the following
Self-Test.

Self-Test

24. What crystal type is observed for

(a) NaCl

(b) SO_2 _____

(c) Ni   _____

(d) $MgCl_2$ _____

(e) SiC   _____

25. CO_2 forms soft crystals (dry ice) that sublime (evaporate) at -78°C. SiO_2, on the other hand, forms hard high melting crystals (sand). What crystal type does each of these form?

_____

26. UF_6 forms soft crystals that melt at 64.5°C. What is the probable type of crystal formed by UF_6?

_____

27. An element, formerly called columbium, melts at 2468°C, is soft and shiny, and conducts electricity. What is this type of crystal?

_____

New Terms

molecular crystal   covalent crystal
ionic crystal   metallic crystal

8.10 LIQUID CRYSTALS

Objectives

To learn the properties of these substandes that have some properties of both liquids and crystals.

Review

Substances that form liquid crystals have rodlike molecules. The three classes of liquid crystals are nematic, smectic, and cholesteric. Their differences are shown in Figure 8.33.

Self-Test

28. What type of liquid crystal is used in liquid crystal display in wristwatches and calculators?

_____

29. What type of liquid crystal changes color dramatically when its temperature is changed? _____

New Terms

liquid crystal
nematic liquid crystal

smectic liquid crystal
cholesteric liquid crystal

8.11   HEATING AND COOLING CURVES; CHANGES OF STATE

Objectives

To observe what changes take place when heat is gradually added to a solid, or gradually removed from a gas. You should learn the kinds of energy changes that take place along the various segments of a heating or cooling curve.   Learn the definition of supercooling and amorphous solid.

Review

On those portions of a heating or cooling curve where the temperature is changing (the slanted line segments in Figures 8.34 and 8.35) the kinetic energy of the molecules is changing. The horizontal segments, where the temperature remains constant, correspond to changes in potential energy during a phase change (gas $\longleftrightarrow$ liquid, liquid $\longleftrightarrow$ vapor).

Supercooling is a phenomenon that occurs when a liquid is cooled so rapidly that its temperature drops below the ordinary freezing point before the molecules have an opportunity to assemble themselves into the proper arrangement to form a crystal. Amorphous solids such as glass are actually supercooled liquids.

Self-Test

30. Refer to the cooling curve which follows to answer the following:
    (a) Which line segment corresponds to the conversion of vapor to liquid? _____

(b) Which line segments correspond to changes in kinetic energy?

_____

(c) Which line segments correspond to changes in potential energy?

_____

(d) Use a dotted line to indicate the effect of supercooling.

(e) What is the boiling point of the substance? _____

(f) What is the melting point of the substance? _____

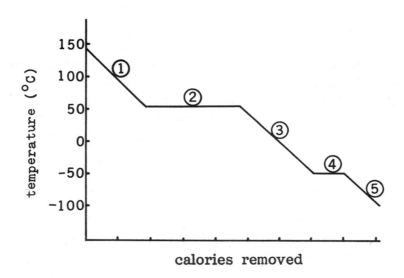

New Terms

heating curve   supercooling
cooling curve   amorphous solid

8.12 VAPOR PRESSURE OF SOLIDS

Objectives

To learn how solids, like liquids, are able to evaporate by sublimation.

Review

     Sublimation is the direct conversion of solid to vapor without passing through the liquid state. If the solid is placed in a closed container, the vapor can come to equilibrium with the solid. The pressure exerted by the vapor is called the vapor pressure of the solid. It too rises with increasing temperature.

Self-Test

31. The heat of sublimation, ΔH_{subl}, is the energy required to convert one mole of solid directly to one mole of vapor. The value of ΔH_{subl} is always greater than ΔH_{vap}. Why is this so?

_____

_____

32. Which substance would have a higher vapor pressure at -90°C, solid H_2O or solid CO_2? Why? _____

_____

New Terms

equilibrium vapor pressure of a solid

8.13 PHASE DIAGRAMS

Objectives

     To learn how a phase diagram can be used to define the limits of temperature and pressure over which the different states of a substance can exist.

Review

     The type of phase diagram discussed in this section contains three lines that define pressures and temperatures at which equilibria can exist between two phases. The solid-vapor equilibrium line is the vapor pressure curve for a solid. The liquid-vapor line, which terminates at the critical temperature and pressure, is the vapor pressure curve for the liquid. The solid-liquid line gives the melting point at different pressures. These lines serve as boundaries to temperature/pressure regions where only one phase can exist. Review Figure 8.38 and learn which regions of the diagram correspond to solid, liquid, and vapor. Remember that the three equilibrium lines intersect at the triple point - the temperature and pressure at which all three states can coexist in dynamic equilibrium.

The relationships between temperature, pressure and phase changes are covered in detail in Figures 8.39 to 8.41. Review this material, too, so you can follow the changes that take place moving either horizontally or vertically on a phase diagram.

In most substances the solid-liquid line slants to the right. This is because in most cases the solid is more dense than the liquid. Application of pressure on a liquid (moving up vertically at constant temperature) converts it to the more dense (more compact) solid phase. You should be able to make this prediction on the basis of Le Châtelier's principle.

Self-Test

33. Sketch a phase diagram for a substance whose triple point occurs at $-10°C$, 25 torr and whose normal melting and boiling points are $-5°C$ and $120°C$, respectively. Identify the solid-liquid (S-L), liquid-vapor (L-V) and solid-vapor (S-V) lines. Indicate the solid, liquid, and vapor regions.

New Terms

triple point
phase diagram

Answers to Self-Test Questions

1. b   2. d   3.(a) alcohol (b) alcohol 4. d   5. It lowers the surface tension of the liquid.   6.(a) dipole-dipole and London (b) London   (c) hydrogen bonding and London   (d) dipole-dipole and London   (e) hydrogen bonding and London
7. London forces are intermittent because the dipoles come and go, existing only briefly.   8. $SiH_4 < C_2H_6 < HCl < H_2S < NH_3 <$ HF (arranged in order of increasing ΔH_{vap})
9.(a) AsH_3 (b) SiH_4 (c) H_2O 10.(a) vapor   (b) liquid 11. b
12. HCl 13.(a) no (b) they are large   14. Because HF can

form only two hydrogen bonds while H_2O can form four.   Four weaker bonds turn out to be stronger than two strong bonds. 15. Increasing pressure raises the boiling point.   16. steam 17. It gives a measure of the difference between the strengths of attractive forces in two phases that each contain relatively strong attractive forces.   18. None. They are identical.   19. 316 pm 20. Two   21. Entirely within the unit cells so that the Cs^+ to Cl^- ratio can be 1:1   22. 291 pm   23.(a) 133 pm   (b) 1.33 Å 24.(a) ionic   (b) molecular   (c) metallic   (d) ionic   (e) covalent 25. CO_2 – molecular; SiO_2 – covalent   26. molecular   27. metallic (the element is now called niobium)   28. nematic   29. cholesteric 30.(a) 2   (b) 1, 3, 5   (c) 2, 4   (d) See Figure 8.36 in the text. (e) 50°C   (f) –50°C   31. The strengths of the attractive forces are greater in the solid than in the liquid.   Therefore, more energy must be added to convert the solid to a gas than to change the liquid to a gas.   32. CO_2, because it has weaker attractive forces (CO_2 is nonpolar, H_2O is polar and exhibits hydrogen bonding).

33.

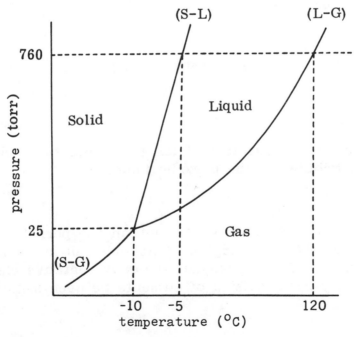

9 THE PERIODIC TABLE REVISITED

In studying the preceding chapters, you have acquired some understanding of the forces between atoms within chemical compounds - the chemical bonds between atoms - and of the forces that exist between molecules - intermolecular attractions. With this as background, we now turn to a discussion of some chemical and physical properties of the elements, with an emphasis on how these properties relate to an element's position within the periodic table.

9.1   METALS, METALLOIDS, AND NONMETALS

Objectives

To learn how elements that fall into these three classes are distributed in the periodic table.

Review

In general, metals are on the left in the periodic table and nonmetals are in the upper right. Metalloids fall in a band from boron to astatine. Looking only at the representative elements, there is a nearly even division of elements between metals and nonmetals.

Self-Test

1. Use the periodic table on the inside front cover of your text-

book rather than Figure 9.1 to answer this question.

(a)   Which of these are metals: Mo, Dy, Ge, Au, Se?

_____

(b)   Which of these are nonmetals: Cd, Ho, I, C, Sb, Ar?

_____

(c)   Which of the elements in parts (a) and (b) are metalloids?

_____

2.   Among the representative elements:
(a)   How many metals are there?   _____

(b)   How many nonmetals are there?   _____

New Terms

9.2   PHYSICAL PROPERTIES OF METALS

Objectives

To learn which physical properties serve to identify metals.

Review

The metallic lattice results because metal atoms have few valence electrons which are loosely held. Metals are characterized by having good electrical and thermal conductivities, a "metallic luster," and by being ductile and malleable.

The melting points of metals range from very high to quite low. Those with the highest melting points follow lanthanum in period 6. In general, the hardness of metals roughly parallels their melting points - high-melting metals tend to be hard, and vice versa. Melting point and hardness reflect the strengths of the attractive forces between positive ions in the metallic lattice and their surrounding "electron sea." Among the transition metals there is evidence of some covalent bonding between atoms as well.

Self-Test

3. Define:
    (a) ductility _____

    (b) malleability _____

4. Which element is used as a filament in light bulbs? _____

    Why? _____

5. Potassium has a melting point of 63.7°C. Would you expect potassium metal to be hard or soft?

    _____

6. What evidence is there for covalent bonding among the transition metals?

    _____

7. How do you expect the melting point of aluminum to compare with that of magnesium? Explain.

    _____

    _____

    _____

New Terms

malleability
ductility

9.3   CHEMICAL PROPERTIES OF METALS

Objectives

    To learn some general chemical properties of metals.

Review

    When metallic elements react, they become oxidized.
(There are a few exceptions, but you need not worry about them now.) The corrosion of a metal by reaction with oxygen is an example. The most reactive metals are found in Group IA. The least reactive occur in period 6 of the periodic table, as shown in Figure 9.8. The unreactive metals are called noble metals. In general, trends in the ease of oxidation of metals parallels

trends in ionization energy - metals with low ionization energies tend to be easily oxidized.

The ease of oxidation of a metal can be judged by its tendency to react with <u>nonoxidizing acids</u> - acids in which the only oxidizing agent is H^+. Examples are HCl and H_2SO_4. Those that react rapidly with acids are easily oxidized. The general reaction is

$$\text{metal + acid} \longrightarrow H_2 + \text{salt}$$

For example,

$$Fe + 2HCl \longrightarrow H_2 + FeCl_2$$

Metals of Group IA and those from Ca to Ra in Group IIA are so reactive that they produce hydrogen by reaction with water. Other metals (copper, for instance) are unreactive toward nonoxidizing acids, but dissolve in the oxidizing acid, HNO_3. The noble metals will only dissolve in aqua regia.

Metals of Group IA and IIA produce ions with charges of 1+ and 2+, respectively. Aluminum produces Al^{3+} when it reacts. Heavier metals in Groups III, IV and V form two oxidation states - one corresponding to the loss of the p electron(s) from the valence shell, and the other corresponding to the loss of all the electrons from the valence shell. The lower oxidation states becomes progressively more stable relative to the higher ones as we go down a group. Many of the transition metals have a 2+ oxidation state corresponding to the loss of the outer pair of s electrons.

Self-Test

Refer to the periodic table on the inside front cover of the text to answer the following questions.

8. Which element in each pair is more easily oxidized?

(a) Rb or Sr   _____

(b) Ti or Ir   _____

(c) Ca or Fe   _____

9. Write an equation for the reaction between hydrochloric acid and
(a) magnesium   _____

(b) aluminum   _____

   (c) nickel (to give Ni^{2+}) _____

10. Write an equation for the reaction of water with

   (a) potassium   _____

   (b) strontium   _____

11. Name two noble metals. _____

12. What is the nitrogen containing product when copper reacts
     with
   (a) concentrated HNO_3 _____

   (b) dilute HNO_3 _____

13. Does silver react with hydrochloric acid?   _____

14. Write an equation for the reaction of silver with concentrated
     nitric acid. (Silver is oxidized to the 1+ state.)

     _____

15. What is aqua regia? _____

16. If you had to make an "educated guess," what would you
     predict one of the oxidation states of chromium to be?

     _____

New Terms

corrosion   aqua regia
noble metal

9.4   TRENDS IN METALLIC BEHAVIOR

Objectives

   To learn how the metallic character of the elements varies
   within the periodic table.

Review

   Trends in metallic character parallel trends in the electronegativities of the elements - elements having high electronegativities are less metallic than those with low electronegativities.
(Elements with low electronegativity are said to be electropositive.)

Metallic character increases from top to bottom in a group and decreases from left to right across a period.

Acidic or basic properties of oxides serve as a measure of metallic character. Metal oxides are basic and nonmetal oxides are acidic. Some oxides (e.g., BeO and Al_2O_3) are amphoteric. Aluminum itself is amphoteric and dissolves in both acid and base with the liberation of hydrogen.

Self-Test

17. Choose the element with the more metallic character in each set below:

 (a) Mg or Sr   _____

 (b) Al or Mg   _____

 (c) Ga, Cs, Tl, or Ba   _____

18. Which oxide is more basic, Al_2O_3 or In_2O_3?   _____

19. Which oxide is more acidic, Li_2O or BeO?   _____

20. Which is the most electropositive element among Mg, K, Sn, Cs?

          _____

21. Oven cleaners often contain NaOH. Why shouldn't they be used to clean aluminum oven pans? `_____

   _____

22. Write an equation for the reaction of calcium oxide with hydrochloric acid. _____

23. Which group in the periodic table best illustrates the variation in metallic character within a group?   _____

New Terms

electropositive
amphoteric

9.5   IONIC-COVALENT CHARACTER OF
METAL-NONMETAL BONDS

Objectives

To learn how the relative degree of covalent character of metal-nonmetal bonds can be explained in terms of cation charge and size.

Review

The degree to which a cation is able to polarize the electron cloud of an anion is related to the cation's ionic potential, ϕ. Ionic potential is the ratio of an ion's charge to its radius. Recalling from Chapter 3 the way in which atomic and ionic size varies within the periodic table, you can predict the relative degree of covalent character to metal-nonmetal bonds, as discussed on Page 297.

Self-Test

24. Arrange the following ions in order of increasing ionic potential:

	ion	radius ($\overset{\circ}{A}$)
(a)	Er^{3+}	0.96
(b)	Be^{2+}	0.39
(c)	Ti^{4+}	0.68

25. In each pair below, choose the substance with the greater degree of covalent character.

(a) $MgCl_2$ or $BeCl_2$ _____

(b) Al_2S_3 or Ga_2S_3 _____

(c) CrO_2^- or CrO_4^{2-} _____

(d) $FeCl_2$ or $FeCl_3$ _____

(e) SnS_2 or SnS _____

(f) CaS or Ga_2S_3 _____

New Terms

ionic potential

9.6   COLORS OF METAL COMPOUNDS

Objectives

To learn what gives rise to the observed color of compounds and to observe how color can be related to the degree of covalent character in certain metal-nonmetal compounds.

Review

The color that we observe for a particular compound is determined by the colors that the compound absorbs from white light. We see the colors that are not absorbed.

Absorption of light by typical ionic compounds is of the charge transfer type. As the metal-nonmetal bond becomes more covalent, less energy is needed to cause the charge transfer and the absorption band shifts from the UV region toward the blue end of the visible spectrum. This causes the compound to appear colored. The intensity of the observed color is proportional to the degree of covalent character.

The degree of covalent character of a metal-nonmetal bond increases with the size of the anion and with the charge on the anion. Both factors make the anion more polarizable.

Self-Test

26. Which is more covalent, Na_2S (Na—S distance = 283 pm) or NaCl (Na—Cl distance = 281 pm)?

                          _____

27. Choose the member of each pair that would be most deeply colored.

(a) Ag_2O or Ag_2S _____

(b) ZnS or ZnSe   _____

New Terms

charge transfer process
charge transfer absorption band

9.7   SOME PHYSICAL PROPERTIES OF NONMETALS AND METALLOIDS

Objectives

To learn some of the physical properties that distinguish nonmetals and metalloids from metals.

Review

The nonmetals and metalloids occur in a variety of forms at room temperature. Those in Group 0 occur as monatomic gases. Other gaseous nonmetals are diatomic (H_2, N_2, O_2, F_2, and Cl_2). Also diatomic are Br_2 (a liquid) and I_2 (a solid). The remainder of the nonmetals (and the metalloids) are solids of varying degrees of complexity. Except for the noble gases, nonmetal atoms complete their valence shells by covalent bonding, so no free electrons are available for metallic conduction. Nonmetals are therefore poor conductors of both heat and electricity. Nonmetallic solids are brittle and lack metallic luster. A special property of metalloids is their semiconductivity.

Self-Test
---

28. How do the melting points of the halogens vary?

     _____

29. Which nonmetals are solids at room temperature?

     _____

New Terms

9.8   CONDUCTORS, NONCONDUCTORS, AND SEMICONDUCTORS

Objectives

To understand how metals conduct electricity. You should also learn what distinguishes an insulator from a conductor and why some substances behave as semiconductors.

Review

In a solid, atomic orbitals of the same or similar energies on all of the atoms in the crystal combine to form energy bands. The band formed from the valence shell orbitals is termed the valence band. All bands below the valence band are completely occupied by electrons, which are held tightly to their respective atoms. A band that is empty or partially filled is called a conduction band. The conduction band extends continuously throughout the solid and electrons in the conduction band move freely from atom to atom.

A conductor is a substance that has either a partially filled conduction band (like Na) or a conduction band that overlaps the valence band (like Mg). In an insulator there's a large energy gap between a filled valence band and the empty conduction band, and it is difficult to raise an electron into the conduction band. A semiconductor has a small gap between the valence band and conduction band and thermal energy is sufficient to raise some electrons to the conduction band. Increasing the temperature increases the population of the conduction band and increases the conductivity. Addition of an impurity alters the conductivity of a semiconductor. If the impurity has fewer electrons than atoms of the host lattice, a p-type semiconductor occurs in which charge is transferred by the movement of positively charged "holes."

When the impurity has more electrons than atoms of the host lattice, an n-type semiconductor is produced in which charge is transported by the extra electrons supplied by the guest atoms.

Self-Test

30. Why are nonmetals nonconductors of electricity? _____

_____

31. What type of semiconductor occurs if a small quantity of

    (a) Ga is added to pure As? _____

    (b) Se is added to pure As? _____

    (c) P is added to pure Si? _____

New Terms

band theory of solids   p-type semiconductor
valence band   n-type semiconductor
conduction band

9.9   MOLECULAR STRUCTURES OF
THE NONMETALS AND METALLOIDS

Objectives

   To look for generalizations that can be applied to understanding the degrees of complexity of the structures of
the elemental nonmetals and metalloids.

Review

   The key to understanding the molecular structures of
these elements is the idea that elements in period 2 are able to
form strong π-bonds while elements in the following periods form
much weaker π-bonds. As a result, double and triple bonds are
observed among period 2 elements, but period 3 elements tend to
favor two or three single bonds to separate atoms, rather than
π-bonds to just one other atom.

   The following are the most important items to review:

   (1) allotropism involving oxygen and carbon
   (2) the structure of graphite, which accounts for
        its electrical conductivity
   (3) the structure of the S_8 ring in sulfur
   (4) the structure of white phosphorus
   (5) the unusual structure of boron

Self-Test

32. How many atoms do the following elements bond to in their
     elemental forms?

     (a) Cl _____   (c) P   _____

     (b) S   _____   (d) Si _____

33. How does graphite differ from diamond? _____

_____

34. What is the structure of elemental sulfur? _____

_____

35. What is the structure of white phosphorus? _____

_____

36. What two physical properties are the result of the unusual structure of boron?

_____

_____

New Terms

allotropism   red phosphorus
white phosphorus

9.10 CHEMICAL PROPERTIES OF THE NONMETALS AND METALLOIDS

Objectives

To learn some chemical properties of nonmetals and how they vary within the periodic table, and to understand why some acids are stronger than others. You should be able to predict the relative acid strengths among oxo-acids (acids containing hydrogen, a nonmetal, and oxygen) and among hydro acids (acids containing only hydrogen and a nonmetal).

Review

The nonmetals and metalloids (referred to in this section as nonmetals, without making a distinction) form more compounds than the metals because they form covalently bonded molecules as well as ions. The oxides of nonmetals are acidic as are the water solutions of many of their hydrogen compounds (e.g., HCl, HBr).

Below are summarized the trends in acidity that we observe among different kinds of acids. You should be aware of these, and the Self-Test provides a way to test your ability to apply the rules. In addition, you should try to understand why these trends occur. If you're not sure about the reasons, reread the explanations provided in the text.

Oxoacids, H_nXO_m

(1) Strength increases as the number of lone oxygen atoms increases.

(2) For acids of the same general formula, strength decreases as the central atom gets larger (as you descend a group).

(3) In general, attaching an electronegative atom (or group of atoms) to an atom bonded to an O—H group increases the acidity of the O—H group.

(4) An increase in the oxidation number of an atom that is bonded to an O—H group increases the acidity of the O—H group.

Hydro acids (binary acids), H_nX

(1) Strength increases from left to right within a period.

(2) Strength increases from top to bottom within a group.

Self-Test

37. In each pair below, choose the strongest acid.

(a) H_2SeO_3, H_2SeO_4 _____

(b) $HBrO_3$, $HBrO$ _____

(c) HNO_3, HNO_2 _____

(d) H_3PO_4, H_2SO_4 _____

(e) H_3PO_4, H_3AsO_4 _____

(f) H_3N, H_2O _____

(g) PH_3, NH_3 _____

(h) H_2Se, HBr _____

New Terms

Answers to Self-Test Questions

1.(a) Mo, Dy, Au (b) I, C, Ar (c) Ge, Sb 2.(a) 18 (b) 17
3.(a) ability to be drawn or stretched into wires (b) ability to
be hammered into thin sheets 4. Tungsten, because it has such
a very high melting point and conducts electricity. 5. soft
6. Melting points increase with increasing number of unpaired
electrons. 7. Al should be higher because of the Al^{3+} ion in its
metallic lattice (actually, it is only slightly higher). 8.(a) Rb
(b) Ti (c) Ca 9.(a) $Mg + 2HCl \longrightarrow MgCl_2 + H_2$
(b) $2Al + 6HCl \longrightarrow 2AlCl_3 + 3H_2$ (c) $Ni + 2HCl \longrightarrow NiCl_2$
10.(a) $2K + 2H_2O \longrightarrow 2KOH + H_2$
   (b) $Sr + 2H_2O \longrightarrow Sr(OH)_2 + H_2$
11. platinum and gold 12.(a) NO_2 (b) NO 13. no
14. $Ag + 2HNO_3 \longrightarrow AgNO_3 + NO_2 + H_2O$
15. 1 part HNO_3, 3 parts HCl (by volume) 16. 2+ (a common
oxidation state among the transition metals) 17.(a) Sr (b) Mg
(c) Cs 18. In_2O_3 19. BeO 20. Cs 21. Al reacts with base
(see Page 296) 22. $CaO + 2HCl \longrightarrow CaCl_2 + H_2O$ 23. Group IVA
24. $Er^{3+} < Ti^{4+} < Be^{2+}$ 25.(a) $BeCl_2$ (b) Al_2S_3 (c) CrO_4^{2-}
(d) $FeCl_3$ (e) SnS_2 (f) Ga_2S_3 26. Na_2S 27.(a) Ag_2S
(b) SnSe 28. $F_2 < Cl_2 < Br_2 < I_2$ 29. C, P, S, Se, I
30. All the electrons are localized in covalent bonds.
31.(a) p-type (b) n-type (c) n-type 32.(a) one (b) two
(c) three (d) four 33. graphite - planar sheets held to each
other by London forces; diamond - three-dimensional covalent
structure 34. Puckered S_8 ring (Figure 9.21) 35. Tetrahedral
P_4 molecule (Figure 9.22) 36. hardness and high melting point
37.(a) H_2SeO_4 (b) $HBrO_3$ (c) HNO_3 (d) H_2SO_4 (e) H_3PO_4
(f) H_2O (g) PH_3 (h) HBr

10 PROPERTIES OF SOLUTIONS

In Chapter 6 solutions were discussed in terms of their usefulness as a medium for carrying out chemical reactions. The focus of this chapter is on the physical properties of solutions (boiling point, freezing point, vapor pressure, etc.) as they are affected by the presence of a solute.

10.1   TYPES OF SOLUTIONS

Objectives

To review the different kinds of solutions that can be formed.

Review

Solutions can be formed between: gas-gas (gaseous solutions); liquid-gas, liquid-liquid, liquid-solid (liquid solutions); solid-gas, solid-liquid, solid-solid (solid solutions).

Solid solutions can be of two types, substitutional and interstitial. Review the meanings of these.

Self-Test

1.   What kind of solutions do the following represent?

(a)   black coffee (without sugar)   _____

(b)   brass   _____

    (c)   a carbonated beverage   _____

    (d)   auto exhaust   _____

    (e)   a martini (without the olive)   _____

New Terms

alloy   interstitial solid solution
substitutional solid solution

10.2   CONCENTRATION UNITS

Objectives

To define additional concentration units that are useful in treating the physical properties of solutions. You should learn to convert between the different units.

Review

Be sure you know the definitions of the following units. The most common reason that students have difficulty handling concentration units is because they fail to learn the definitions.

mole fraction
$$X_A = \frac{n_A}{n_A + n_B + n_C + \cdots}$$

n_A, n_B, n_C, etc. are the number of moles of each component of the solution. Mole percent (mol percent) equals 100 x mole fraction.

weight fraction
$$w_A = \frac{\text{weight of A}}{\text{total weight of solution}}$$

Weight percent, of course, equals 100 x weight fraction,
$$\% \ A \ (w/w) = 100 \times w_A$$

molarity
$$M = \frac{\text{moles of solute}}{\text{liters of solution}}$$

normality
$$N = \frac{\text{equivalents of solute}}{\text{liters of solution}}$$

$$\underline{\text{molality}} \qquad m = \frac{\text{moles of solute}}{\text{kilograms of solvent}}$$

(Be careful to clearly distinguish between molarity and molality; they sound and look alike but are defined quite differently.)

Conversions among molality, mole fraction, and weight fraction are straightforward and require only the molecular weights of solute and solvent (see Examples 10.1 - 10.3 in the text). To convert any of these to molarity (or vice versa) requires the density of the solution, as shown in Example 10.4 and the additional example below.

Example 10.1

An aqueous solution of $CuSO_4$, having a mole fraction of $CuSO_4$ equal to 0.0127, has a density of 1.106 g/ml. What is the molarity of the $CuSO_4$?

Solution

We begin by assuming that we have a total of 1 mol of solute and solvent. Then there are

$$0.0127 \text{ mol } CuSO_4$$
$$0.9873 \text{ mol } H_2O$$

To obtain the volume of the solution, which we need to compute the molarity, we first must calculate the total mass of the solution and then apply the density.

$$0.0127 \text{ mol } CuSO_4 \times \left(\frac{159.5 \text{ g } CuSO_4}{1 \text{ mol } CuSO_4}\right) = 2.03 \text{ g } CuSO_4$$

$$0.9873 \text{ mol } H_2O \times \left(\frac{18.02 \text{ g } H_2O}{1 \text{ mol } H_2O}\right) = 17.79 \text{ g } H_2O$$

$$\text{Total weight of solution} = 19.82 \text{ g}$$

$$\text{volume of solution} = 19.82 \text{ g} \times \frac{1 \text{ ml}}{1.106 \text{ g}} = 17.92 \text{ ml}$$

We now have both the number of moles of $CuSO_4$ and the volume of the solution. The molarity is:

$$\text{molarity} = \frac{0.0127 \text{ mol CuSO}_4}{0.01792 \text{ liter soln.}}$$

$$= 0.709 \text{ M}$$

Self-Test

2. Calculate (a) the molality, (b) mole fraction of solute, and (c) weight fraction of solute in a solution prepared by dissolving 100 g of paraffin (molecular weight 370) in 500 g of benzene, C_6H_6. This solution could be used as a paint remover.

3. A solution of NaBr in water has a weight fraction of NaBr equal to 0.40. What is the (a) mole fraction, (b) mole percent, and (c) molality?

4. The density of the solution in Question 3 is 1.414 g/ml. What is the molarity of the NaBr?

5. A solution of propylene glycol, $C_3H_8O_2$, in water has a molality of 2.65 m. What is the weight percent $C_3H_8O_2$ in the solution?

New Terms

molality

weight fraction

volumetric flask

mole fraction

mole percent

10.3   THE SOLUTION PROCESS IN LIQUID SOLUTIONS

Objectives

To learn what factors control the solubility of substances in liquid solvents.

Review

The key point in this section is that in order for substances to be appreciably soluble in each other, they must possess similar intermolecular attractive forces. Particles that attract each other very strongly tend to congregate and separate from those to which they are weakly attracted.

Remember that when a solute particle is placed in solution it becomes solvated, that is, surrounded by solvent molecules to which it is attracted. When the solvent is water, the term hydration is used.

Self-Test

6. How does hydration of ions help keep them in solution?

_____

_____

7. What does the term "like dissolve like" mean on a molecular level?

_____

_____

New Terms

hydration   miscible
solvation

10.4   HEATS OF SOLUTION

Objectives

To study the energy changes that occur when a solution is formed. You should learn the definition of ideal solution and why some solution processes are exothermic and some are endothermic.

Review

The heat of solution is the energy absorbed or liberated when a solution is formed. In this section you saw that it is possible to divide the total energy change into various contributions, some of which are endothermic and some of which are exothermic.

For liquid solutions formed from a solvent (A) and solute (B), an ideal solution results when the A-B attractions are the same as the A-A and B-B attractions. For an ideal solution, $\Delta H_{soln} = 0$. When the A-B attractions are greater than the A-A and B-B attractions, $\Delta H_{soln} < 0$ and the solution process is exothermic. When the A-B attractions are weaker, $\Delta H_{soln} > 0$ and the solution process is endothermic. These are summarized in Figures 10.6-10.8.

For solutions of solids in liquids the lattice energy (the energy required to separate the solute particles from a crystal) and hydration energy (or solvation energy - the energy released when the solute particle is placed into the solvent cage) must be considered. This is summarized in Figure 10.9 in the text.

Self-Test

8. When acetone (a component of nail polish remover) is dissolved in water, the resulting solution becomes warm. What conclusions can you draw about the relative strength of the attractive forces between acetone and water molecules?

9. Acetone and water (Question 8 above) are completely soluble in each other in all proportions. Does this mean they form ideal solutions?

_____

10. For LiCl, $\Delta H_{soln} < 0$. What does this imply about the relative values of the hydration energy of the ions and the lattice energy of LiCl?

_____

_____

New Terms

heat of solution   hydration energy
ideal solution

10.5   SOLUBILITY AND TEMPERATURE

Objectives

To examine the factors that determine the effect of temperature on solubility, and to learn how substances are purified by fractional crystallization.

Review

A rise in temperature increases solubility if the dissolving of additional solute is endothermic. A fairly good rule of thumb is that the solubility of most solids and liquids in a liquid solvent increases with increasing temperature. The solubilities of gases, however, almost always decrease with increasing temperature.

In the procedure called fractional crystallization, a solid substance containing a soluble impurity is dissolved in a minimum of hot solvent. The solution is cooled and some of the pure desired solid crystallizes, leaving the impurity behind in the solution along with some of the desired material. Even though some of the desired substance is lost, that which is recovered is usually of much higher purity.

Self-Test

11. Why do gases usually become less soluble in liquids as the temperature of the solution is raised?

_____

_____

_____

12. A solid is known to contain 80 g of $NaNO_3$ and 5 g of NaCl. The solubility of NaCl is 36 g/100 g H_2O at 0°C and 40 g/100 g H_2O at 100°C. The solubility of $NaNO_3$ is 73 g/100 g H_2O at 0°C and 180 g/100 g H_2O at 100°C.

   (a) What is the minimum amount of boiling water (100°C) necessary to dissolve all of the solid?

   _____

   (b) If the solution is cooled to 0°C, how much $NaNO_3$ will separate as pure solid?

   _____

New Terms

fractional crystallization

10.6   THE EFFECT OF PRESSURE ON SOLUBILITY

Objectives

To learn how, and under what circumstances, pressure influences solubility.

Review

Pressure has virtually no effect on the solubility of solids or liquids in liquid solvents. The solubilities of gases, however, are very markedly affected by pressure changes. This can be predicted by Le Châtelier's principle, since an increase in pressure favors a decrease in the number of moles of gas (this would tend to bring the pressure back down). Check with your instructor whether he expects you to treat the effect of pressure on the solubility of a gas quantitatively. If so, review the material below on Henry's law.

Henry's law relates the concentration of a dissolved gas, C_g, to its partial pressure, p_g, over the solution.

$$C_g = k_g p_g$$

k_g is the Henry's law constant.

Self-Test

13. Why are the solubilities of solids in liquids virtually unaffected by pressure?

_____

14. Calculate the solubility of a gas, X, in water at 20°C if its partial pressure is 720 torr ($k = 3.5 \times 10^{-3}$ g/liter torr at 20°C).

_____

New Terms

Henry's law
Henry's law constant

10.7   VAPOR PRESSURES OF SOLUTIONS

Objectives

To learn how the vapor pressure of a solution depends on the relative amounts of solute and solvent.

Review

When a nonvolatile, nondissociating solute is dissolved in a solvent, the vapor pressure of the solvent is diminished because a portion of the surface becomes occupied by molecules unable to enter the vapor phase. Raoult's law relates the vapor pressure to the mole fraction of the solvent in the solution,

$$P_{solution} = X_{solvent}\, P^{\circ}_{solvent}$$

where $P^{\circ}_{solvent}$ is the vapor pressure of the pure solvent.

When two volatile liquids are mixed, the vapor above the solution contains molecules of both. The vapor pressure of the solution is the sum of the partial pressures of each substance. The partial pressures are also determined by Raoult's law. For some substance, A, its partial pressure is

$$P_A = X_A P^{\circ}_A$$

Deviations from Raoult's law occur when the solution is nonideal (recall the definition of an ideal solution in Section 10.4). When ΔH_{soln} is negative, meaning heat is evolved as the solution is formed, the actual partial pressure of each component is less than that calculated from Raoult's law. The reason for this is that the solute and solvent are held more tightly in the solution than in either pure substance. These extra strong solute-solvent attractive forces are responsible for both a negative ΔH_{soln} and negative deviations from Raoult's law. Positive deviations occur when ΔH_{soln} is positive. In this case the A-B attractive forces are less than either A-A or B-B attractions. Molecules are held less tightly than predicted for an ideal solution and the vapor pressure is greater than that calculated from Raoult's law.

Self-Test

15. The vapor pressure of water at 100°C is 760 torr. What is

the vapor pressure of a solution of 200 g of sugar, $C_{12}H_{22}O_{11}$, in 1000 g of H_2O at this same temperature? Will the solution boil at 100°C under an atmospheric pressure of 760 torr?

16. At 85°C, ethylene bromide ($C_2H_4Br_2$) has a vapor pressure of 170 torr and propylene bromide ($C_3H_6Br_2$) has a vapor pressure of 127 torr. These substances form very nearly an ideal solution. What is the vapor pressure of a solution containing 100 g of each?

New Terms

Raoult's law
positive deviations from Raoult's law
negative deviations from Raoult's law

10.8 FRACTIONAL DISTILLATION

Objectives

To see how a distillation process can often be used to separate mixtures of volatile liquids.

Review

Remember that at the boiling point the sum of the partial pressures of the components above a mixture equals the atmospheric pressure.

$$P_{atm} = p_A + p_B = P_{Total}$$

The partial pressures, p_A and p_B, in turn, are found by Raoult's law.

$$p_A = X_A P_A^\circ$$

Remember that with Raoult's law X_A and X_B are the mole fractions of A and B <u>in the liquid</u>.

In the vapor the mole fraction is found from Dalton's law. For example,

$$p_A = X_A P_{Total}$$

which gives

$$X_A = \frac{p_A}{P_{Total}} = \frac{p_A}{p_A + p_B}$$

Be careful not to confuse the mole fraction in the liquid with the mole fraction in the vapor.

When a liquid mixture boils, the vapor always contains a larger proportion of the more volatile component than does the liquid. Remember this, because it can help you see when you've made a mistake in a calculation.

A boiling point diagram is shown in Figure 10.16 in the text. Remember that the upper curve gives the composition of the vapor; the lower curve gives the composition of the liquid. The compositions of vapor and liquid in equilibrium are connected by a tie line. Review how repeated boiling and condensation of the vapor gradually gives a liquid richer in the more volatile component.

Mixtures having large deviations from Raoult's law form azeotropes. These mixtures have either maxima or minima in their boiling point curves. They can only be separated into one pure component plus the liquid mixture having the composition at the maximum or minimum of the curve.

Self-Test

17.   Two substances, A and B, have vapor pressures at 85°C of 800 torr and 300 torr, respectively. What will be the composition of a mixture of these substances that boils at 85°C under 1 atm pressure? (Hint - use Raoult's law and remember that $X_B = 1 - X_A$) _____

18.   If mixtures of two liquids show large positive deviations from Raoult's law, do they form a maximum boiling azeotrope or a minimum boiling azeotrope?

_____

New Terms

fractional distillation   tie line
boiling point diagram   azeotrope

10.9   COLLIGATIVE PROPERTIES OF SOLUTIONS

Objectives

   To see how the vapor pressure lowering of a solution by a
nonvolatile solute causes a boiling point elevation and
freezing point depression.   You should learn how this
phenomenon permits the determination of molecular weights.
In addition, you should learn how the dissociation of an
electrolyte produces abnormally large changes in boiling
and freezing points.

Review

   You should examine Figure 10.18 in the text to be sure
you understand why a nonvolatile solute raises the boiling point
and lowers the freezing point.

   Remember that

$$\Delta T_b = K_b m$$

and $$\quad \Delta T_f = K_f m$$

where m is the molality - the number of moles of solute particles
per 1000 g (1 kilogram) of solvent.   The specific values of K_b
and K_f depend on the solvent (Table 10.5).

   These relationships are useful in two ways.   Knowing m
and K_b or K_f, you can calculate the changes in boiling and
freezing points.   This might be important, for example, if you
wanted to know the properties of an antifreeze solution.   The
important application to chemistry is in the determination of
molecular weights.   Here we measure ΔT, and knowing K we can
calculate the molality.   From a knowledge of the weights of solute
and solvent we can obtain a relationship between weight and number of moles, from which the molecular weight can easily be computed.

Example 10.2

   5.48 g of a solid are dissolved in 200 g of water to give a
solution having a freezing point of -0.850°C. What is the molec-

ular weight of the substance?

Solution

From the freezing point depression, 0.850°C, we can calculate the molality.

$$m = \frac{\Delta T}{K} = \frac{0.850°C}{1.86°C/molal} = 0.457 \text{ molal}$$

This translates to the ratio,

$$\frac{0.457 \text{ mol solid}}{1.00 \text{ kg } H_2O} \tag{1}$$

Next we calculate the ratio of mass of solid to kilograms of water.

$$\text{ratio} = \frac{5.48 \text{ g solid}}{0.200 \text{ kg } H_2O}$$

Dividing numerator and denominator by 0.200 we get,

$$\text{ratio} = \frac{27.4 \text{ g solid}}{1.00 \text{ kg } H_2O} \tag{2}$$

Ratio (1) must equal ratio (2) because we are dealing with the same solution. Since their denominators are the same, their numerators must be equal. The next step then is to equate numerators.

$$0.457 \text{ mol solid} = 27.4 \text{ g solid}$$

Finally, divide through by 0.457 to get the weight of one mole.

$$1 \text{ mol solid} = 60.0 \text{ g}$$

The molecular weight, therefore, is 60.0 g/mol.

Self-Test

19. Calculate the boiling point elevation and the actual boiling point of a solution of 35.0 g of a solute having a molecular weight of 210 in 450 g of benzene. Use the data in Table 10.5.

20. What is the molecular weight of an unknown substance if a solution of 0.00213 g X in 0.100 g of camphor has a freezing point of 174.5°C? Use the data in Table 10.5.

21. Calculate the freezing point depression produced by a solution of 0.100 g of a compound having a molecular weight of 10,000 dissolved in 100 g of water.

_____

New Terms

colligative properties
molal boiling point elevation constant
molal freezing point depression constant

10.10   OSMOTIC PRESSURE

Objectives

To see how osmosis provides a means of determining very large molecular weights.

Review

If you worked through the solution to Question 21 above, you saw that when the molecular weight is large, ΔT_f is very small. This makes it nearly impossible to measure. Measurements of osmotic pressure provide means of calculating very high molecular weights because the osmotic pressure produced by even very dilute solutions is measurable.

The van't Hoff equation is easy to remember - it looks like the ideal gas law.

$$\pi V = nRT$$

If you know π, V, R, and T, you can calculate the number of moles of solute in the solution.

_____

Example 10.4

A 0.010-g sample of starch in 5.0 ml of water at 25°C produces a solution having an osmotic pressure of 2.3 torr. What is the average molecular weight of the starch molecules?

Solution

First, let's express π in atm.

$$\pi = 2.3 \text{ torr } \times \left(\frac{1 \text{ atm}}{760 \text{ torr}}\right) = 3.0 \times 10^{-3} \text{ atm}$$

We also have

$$V = 0.005 \text{ liter}$$
$$R = 0.0821 \text{ liter atm/mol K}$$
$$T = 298 \text{ K}$$

Solving the van't Hoff equation for n gives

$$n = \frac{\pi V}{RT}$$

and substituting,

$$n = \frac{(3.0 \times 10^{-3} \text{ atm})(0.005 \text{ liter})}{(0.0821 \text{ liter atm/mol K})(298 \text{ K})}$$

$$n = 6.1 \times 10^{-7} \text{ mol}$$

From the quantity of starch placed in the solution we have

$$6.1 \times 10^{-7} \text{ mol} = 0.010 \text{ g}$$

$$1 \text{ mol} = 1.6 \times 10^{4} \text{ g} = 16,000 \text{ g}$$

The molecular weight is 16,000.

Self-Test

22. A solution of a water-soluble polymer (0.20 g) in 10 ml of water at 20°C has an osmotic pressure of 0.10 torr. What is the molecular weight of the polymer?

New Terms

osmosis   osmotic pressure
semipermeable membrane   isotonic
dialysis

10.11 SOLUTIONS TO ELECTROLYTES

Objectives

To learn how the dissociation of an electrolyte in an aqueous solution affects the colligative properties. You should also learn that in solutions of electrolytes the ions are not totally independent of each other.

Review

Electrolytes dissociate to produce more moles of particles than moles of solute. One mole of NaCl produces 2 mol of particles. The freezing point and boiling point changes are therefore larger than they would be if no dissociation occurred. To obtain the actual ΔT, multiply the ΔT calculated using the molal concentration of salt by the number of ions produced when one formula unit dissociates. For example, for a 1.00 m solution of $CaCl_2$ we would calculate $\Delta T = 1.86°C$. Since $CaCl_2$ produces three ions per $CaCl_2$ formula unit, the actual ΔT is three times as large as we originally calculated.

$$\Delta T_{actual} = 3(\Delta T_{calculated})$$

$$\Delta T_{actual} = 3(1.86°C) = 5.58°C$$

Weak electrolytes produce freezing point depressions and boiling point elevations that are intermediate between those calculated for a nonelectrolyte and those calculated for strong electrolytes. Association of solute particles - the coming together of two or more solute particles in the solution - produces fewer particles and therefore smaller ΔT_f and ΔT_b than expected.

As the concentration of an electrolyte increases, its ions influence each other to a greater extent and are less independent. One way that this can happen is by the formation of ion pairs - groups of oppositely charged ions. As the concentration of ions increases, there is a greater chance for ions of opposite charge to encounter one another and an equilibrium concentration of ion pairs can be created. This, in effect, reduces the number of independent particles that are available to alter the properties of the solution.

The van't Hoff factor, i, is the ratio of the observed freezing point depression (or boiling point elevation) to the freezing point depression (or boiling point elevation) that the substance would exhibit if it were a nonelectrolyte.

Self-Test

23. What value of ΔT_f do you expect for 0.100 m solutions of:

   (a) NaCl _____   (b) $Al_2(SO_4)_3$ _____

24. What is the limiting i factor (at infinite dilution) for the following:

   (a) $MgSO_4$ _____   (c) $Al_2(SO_4)_3$ _____

   (b) K_2SO_4 _____

New Terms

association
dimer

van't Hoff i factor

Answers to Self-Test Questions

1.(a) liquid-solid (b) solid-solid (c) liquid-gas (d) gas-gas
(e) liquid-liquid 2.(a) 0.541 m (b) 0.04 (c) 0.166 3.(a) 0.104
(b) 10.4 (c) 6.48 m 4. 5.50 M 5. 20.2% w/w $C_3H_8O_2$
6. The polar water molecules help shield ions of opposite charge from each other. 7. Substances tend to be soluble in each other only if they have about equal intermolecular attractive forces.
8. Acetone attracts water more than acetone attracts acetone or water attracts water. 9. No. In Question 8 you saw that $\Delta H_{soln} < 0$. For an ideal solution $\Delta H_{soln} = 0$. 10. The hydration energies of the ions must be greater than the lattice energy.
11. Because ΔH_{soln} for a gas is usually negative (exothermic). Increasing the temperature requires adding heat. Le Châtelier's principle predicts that an endothermic change should occur which requires that gas leave the liquid. 12.(a) 44.4 g H_2O (b) 48 g. Note that all of the NaCl is soluble in 44.4 g H_2O. 13. Because they are incompressible. 14. 2.52 g/liter 15. P_{H_2O} = 752 torr. No, the vapor pressure is less than atmospheric pressure.

16. 149.3 torr

17. <u>Solution</u>

$$p_A + p_B = 760$$

$$X_A P_A^o + X_B P_B^o = 760$$

$$X_A P_A^o + (1 - X_A)P_B^o = 760$$

$$X_A(800) + (1 - X_A)(300) = 760$$

$$800X_A + 300 - 300X_A = 760$$

$$500X_A = 760 - 300 = 460$$

$$X_A = 460/500$$

$$X_A = 0.92$$

$$X_B = 1 - X_A = 0.08$$

18. Minimum boiling azeotrope
19. $\Delta T_b = 0.94°C$, $T_b = 80.1 + 0.94 = 81.0° C$
20. 188 g/mol
21. $\Delta T_f = 0.000186°C$
22. 3.7×10^6 g/mol
23. (a) 0.372°C   (b) 0.930°C
24. (a) 2   (b) 3   (c) 5

11 CHEMICAL THERMODYNAMICS

As pointed out in the text, Chapters 11 and 12 deal with the two factors that control whether or not the products of a chemical reaction will form. Thermodynamics controls the feasibility of a reaction in the sense that it determines whether a reaction is possible and how much products can be formed; kinetics (Chapter 12) controls how fast the products are formed.

Thermodynamics deals with energy changes and is applied to physical as well as chemical changes. You will see that a number of thermodynamic principles are developed here using physical systems as examples, followed by the extension of the principles to chemical systems. The applications of thermodynamics range from simple chemical reactions to complex reactions in living organisms.

11.1 SOME COMMONLY USED TERMS

Objectives

To become familiar with some of the terminology to be used in later discussions.

Review

This section defines in a precise way some terms that have been used rather loosely before. It also introduces some new terms. Be sure of their meaning before moving on.

Self-Test

1. Fill in the blanks.

   (a) A process that occurs without a change in temperature is said to be

   _____

   (b) A quantity whose value is independent of the prior history of a sample is called

   _____

   (c) A change that occurs without heat being transferred between the system and its surroundings is said to be

   _____

   (d) The energy required to raise the temperature of 1 g of a substance by 1°C is called

   _____

   (e) The energy required to raise the temperature of the entire system by 1°C is called the
   _____
   of the system.

   (f) The energy required to raise the temperature of one mole of a substance by 1°C is called

   _____

New Terms

system

surroundings

adiabatic

isothermal

state

state function

state variable

equation of state

heat capacity

specific heat

molar heat capacity

11.2   THE FIRST LAW OF THERMODYNAMICS

Objectives

To see how the law of conservation of energy applies to heat transfer and the performing of work. You should also learn what is meant by a reversible process and that the maximum work can only be obtained if a change occurs by a "reversible process."

Review

The first law is simply a restatement of the law of conservation of energy.

$$\Delta E = q - w$$

Remember the following:

$$\Delta E = E_{final} - E_{initial}$$

Also, q is energy (heat) added to the system.

w is energy removed from the system when the system performs work.

ΔE depends only on the initial and final states; E is a state function.

The entire discussion on the isothermal (constant temperature) expansion of an ideal gas was designed to show you that the magnitudes of q and w depend on the way the expansion is carried out - hence q and w are not state functions. Since ΔE is the same regardless of the path between initial and final states, E is a state function.

Remember that one way a system can perform work is to expand against an externally applied pressure. Under a constant opposing pressure,

$$w = P\Delta V$$

If the opposing pressure is zero, no work will be accomplished. This applies not only to gases, but also to chemical systems such as the discharge of the battery described on p 362.

A reversible change is one that takes place in an infinite number of steps, each of which takes place with the opposing force just barely below the driving force for the process. In the reversible expansion of a gas, for instance, the external pressure is initially high and equal to the internal pressure exerted by the gas. As the gas expands the external pressure is dropped at the same rate as the internal pressure drops so that the driving and restraining forces are essentially balanced throughout the entire expansion.

Self-Test

2. A compressed gas (assumed to be ideal) in a cylinder pushes
    back a piston against a constant opposing force of 8.00 atm.
    The initial and final volumes of the gas are 25 ml and 600 ml.
    How much work, expressed in joules, is done by the gas
    during the expansion?

    _____

    Express this answer in calories.   _____

3. Calculate the work done by a gas (in liter atm) as it expands
    from an initial 4 liters at 20 atm to 20 liters at 4 atm

    (a) by a one-step process against an opposing pressure of
         4 atm.

         _____

    (b) By a two-step process in which the opposing pressure
         in the first step is 10 atm and in the second step is
         4 atm.

         _____

New Terms

first law of thermodynamics
internal energy
reversible process

11.3   HEATS OF REACTION: THERMOCHEMISTRY

Objectives

        To express energy changes in chemical reactions in terms
        of thermodynamic quantities.

Review

        For changes at constant volume (and temperature), the
heat of reaction is equal to ΔE.

$$\Delta E = q_v$$

For changes at constant pressure (and temperature), the heat of
reaction is ΔH.

$$\Delta H = q_p$$

H is the enthalpy (also called heat content) and, like E, is a state function. The enthalpy is defined as

$$H = E + PV$$

and at constant pressure

$$\Delta H = \Delta E + P\Delta V$$

A calorimeter is a device used to measure heats of reaction. A bomb calorimeter has a constant volume. When reactions take place in it, the heat liberated is absorbed by the calorimeter and its temperature increases. From the temperature change and a knowledge of the heat capacity of the calorimeter, the amount of heat evolved in the reaction can be calculated (Example 11.2 in the text). The heat of reaction at constant pressure (ΔH) is measured in a similar way, but the contents of the calorimeter are kept at constant pressure.

Usually ΔH and ΔE are computed on a "per mole" basis to make them intensive properties, rather than extensive ones. For a reaction at constant pressure, the difference between ΔH and ΔE depends on the size of the volume change that accompanies the reaction. For reactions involving only liquids and solids, $\Delta H \approx \Delta E$. When gases are consumed or produced, the PV work is given by

$$P\Delta V = \Delta nRT$$

where Δn is the change in the <u>number of moles of gas</u>. Review Example 11.3 in the text.

Self-Test

4. One slice of bread plus sufficient oxygen for complete combustion are placed in a bomb calorimeter having a heat capacity of 36,500 cal/°C. The initial temperature of the calorimeter was 25.00°C. After the combustion was completed, the temperature rose to 26.64°C. How many nutritional Calories (1 Calorie, written with a capital C, equals 1000 calories, or 1 kcal) are contained in one slice of bread, assuming the products of metabolism are the same as the products of combustion?

_____

5. When 0.500 mol of methane, CH_4 (natural gas), is oxidized by oxygen to produce 0.500 mol of CO_2 and 1.00 mol of water vapor, 401 kJ of heat energy are evolved. The bal-

anced equation for the reaction is:

$$CH_4(g) + 2\,O_2(g) \longrightarrow CO_2(g) + 2H_2O(g)$$

(a) What is ΔH in the units kJ/mol CH_4? _____

(b) What is ΔE at 25°C (expressed in the same units)?

_____

6. Methanol may someday replace gasoline as a fuel in automobiles. It burns according to the equation,

$$2CH_3OH(\ell) + 3\,O_2(g) \longrightarrow 2CO_2(g) + 4H_2O(g)$$

Oxidation of 1.000 mol of CH_3OH at 25°C and constant pressure liberates 1280 kJ. What is ΔE for this reaction expressed in kJ/mol?

_____

New Terms

enthalpy
heat content

calorimeter
bomb calorimeter

11.4   HESS'S LAW OF HEAT SUMMATION

Objectives

To use the fact that H is a state function in calculating values of ΔH. You should learn how to combine ΔH values when chemical equations are added or subtracted to produce new equations, as well as the definition of heat of formation.

Review

A chemical equation written to show the energy change that takes place is called a thermochemical equation. Thermochemical equations are always interpreted on a mole basis; they therefore may be written with fractional coefficients. Remember to always indicate the physical state (solid, liquid or gas) of the substances written in a thermochemical equation.

Hess's law says, in effect, that the ΔH for some net reaction is the sum of all of the ΔH's for steps along the way. When thermochemical equations are added together to obtain some final equation, the ΔH for the final equation is the sum of the ΔH's of the thermochemical equations that were combined.

Example 11.1

Add the thermochemical equations below to obtain the value of ΔH for the reaction,

$$2Na_2O_2(s) + 4HCl(g) \longrightarrow 4NaCl(s) + 2H_2O(\ell) + O_2(g)$$

Equations:

$$2Na_2O_2(s) + 2H_2O(\ell) \longrightarrow 4NaOH(s) + O_2(g) \quad ΔH = -30.2 \text{ kcal}$$

$$NaOH(s) + HCl(g) \longrightarrow NaCl(s) + H_2O(\ell) \quad ΔH = -42.8 \text{ kcal}$$

Solution

Inspection of the two equations that are to be combined reveals that we must have 4NaOH on the left to cancel with the 4NaOH on the right when we add the equations. This means the second equation, and its ΔH, must be multiplied by 4 before adding it to the first.

$$2Na_2O_2(s) + 2H_2O(\ell) \longrightarrow 4NaOH(s) + O_2(g) \quad ΔH \doteq -30.2 \text{ kcal}$$

$$4NaOH(s) + 4HCl(g) \longrightarrow 4NaCl(s) + 4H_2O(\ell) \quad \begin{aligned} ΔH &= 4(-42.8 \text{ kcal}) \\ &= -171.2 \text{ kcal} \end{aligned}$$

Adding the equations gives

$$2Na_2O_2(s) + 2H_2O(\ell) + 4NaOH(s) + 4HCl(g) \longrightarrow$$
$$4NaOH(s) + O_2(g) + 4NaCl(s) + 4H_2O(\ell)$$

Adding their ΔH values,

$$ΔH = (-30.2 \text{ kcal}) + (-171.2 \text{ kcal}) = -201.4 \text{ kcal}$$

Remember that in problems of this type, if you have to reverse an equation in order to get cancellation of unwanted formulas, you __must__ also reverse the sign of ΔH.

The heat of formation, $ΔH_f$, is the enthalpy change that occurs when a substance is formed from its elements. Hess's law can be restated in terms of heats of formation as:

$$ΔH_{\text{reaction}} = \left(\begin{array}{c} \text{sum of } ΔH_f \\ \text{of the products} \end{array} \right) - \left(\begin{array}{c} \text{sum of } ΔH_f \\ \text{of the reactants} \end{array} \right)$$

Self-Test

7. Use the equations

$$CaO(s) + SO_3(g) \longrightarrow CaSO_4(s) \qquad \Delta H = -95.9 \text{ kcal}$$

$$Ca(OH)_2(s) \longrightarrow CaO(s) + H_2O(g) \qquad \Delta H = +26.1 \text{ kcal}$$

to obtain the value of ΔH for the reaction,

$$Ca(OH)_2(s) + SO_3(g) \longrightarrow CaSO_4(s) + H_2O(g)$$

_____

8. Use the equations

$$2C_2H_2(g) + 5 O_2(g) \longrightarrow 4CO_2(g) + 2H_2O(g) \quad \Delta H = -2512 \text{ kJ}$$

$$N_2(g) + \tfrac{1}{2}O_2(g) \longrightarrow N_2O(g) \qquad \Delta H = +104 \text{ kJ}$$

to obtain ΔH for the reaction,

$$C_2H_2(g) + 5N_2O(g) \longrightarrow 2CO_2(g) + H_2O(g) + 5N_2(g)$$

_____

9. Use the equations

$$2NO(g) + O_2(g) \longrightarrow 2NO_2(g) \qquad \Delta H = -105 \text{ kJ}$$

$$2N_2O(g) + 3 O_2(g) \longrightarrow 4NO_2(g) \qquad \Delta H = -55 \text{ kJ}$$

$$NO_2(g) + SO_2(g) \longrightarrow NO(g) + SO_3(g) \qquad \Delta H = -47 \text{ kJ}$$

to calculate ΔH for the reaction,

$$2NO(g) + SO_2(g) \longrightarrow N_2O(g) + SO_3(g) \quad \underline{\hspace{3cm}}$$

New Terms

thermochemical equation
Hess's law of heat summation
enthalpy diagram

heat of formation
enthalpy of formation

11.5   STANDARD STATES

Objectives

To establish standard conditions for the comparison of
heats of reaction. You should learn to apply Hess's law
using standard heats of formation.

Review

Standard conditions are chosen to be 25°C (298 K) and 1 atm pressure. A substance in its natural state under these conditions is said to be in its standard state. Standard states are indicated by a superscript zero.

Standard heats of formation, such as those in Table 11.1 in the text, can be used to calculate standard heats of reaction following Hess's law.

$$\Delta H° = (\text{sum of } \Delta H_f° \text{ products}) - (\text{sum of } \Delta H_f° \text{ reactants})$$

This is illustrated in Examples 11.5 and 11.6 in the text. Remember that we always take $\Delta H_f°$ for any pure element in its standard state to be equal to zero.

Sometimes standard heats of formation cannot be measured directly. In these cases a reaction is carried out in which the heat of reaction is measured and the heats of formation of all reactants and products, except the one compound in question, are known. The unknown $\Delta H_f°$ can then be computed. Example 11.7 in the text illustrates this method using a measured heat of combustion.

Self-Test

10. Use the data in Table 11.1 to calculate $\Delta H°$ for the following reactions:

    (a) $CH_4(g) + 4Cl_2(g) \longrightarrow CCl_4(\ell) + 4HCl(g)$ _____

    (b) $Fe_2O_3(s) + 3CO(g) \longrightarrow 3CO_2(g) + 2Fe(s)$ _____

    (c) $C_2H_5OH(\ell) + O_2(g) \longrightarrow CH_3COOH(\ell) + H_2O(g)$

    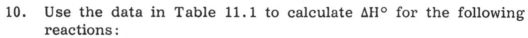

11. The heat of combustion of octane, $C_8H_{18}(\ell)$, a component of gasoline, to produce $CO_2(g)$ and $H_2O(g)$ is -1213.6 kcal/mol of C_8H_{18}. What is the value of ΔH_f of $C_8H_{18}(\ell)$?

    _____

12. From the data in Table 11.1 can you suggest why using N_2O instead of O_2 in a combustion reaction produces a higher flame temperature?

    _____

    _____

New Terms

standard state

11.6   BOND ENERGIES

Objectives

To see how thermodynamic data can be used to obtain information about the strengths of chemical bonds. You should learn how to use bond energies to obtain an estimate of the heat of formation of a compound.

Review

The bond energy is the energy necessary to break a bond to produce neutral fragments. The atomization energy, ΔH_{atm}, is the energy needed to break all of the bonds in a molecule to give neutral atoms.

In many cases the tabulated average bond energies are additive in the sense that they may be used to calculate an atomization energy of a molecule. This can be used as one part of an alternative path from the free elements in their standard states to the compound in question in its standard state.

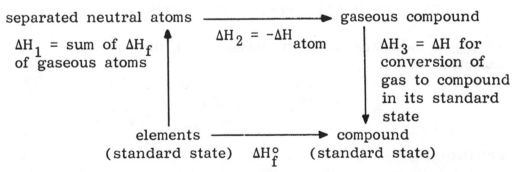

separated neutral atoms ⟶ gaseous compound

ΔH_1 = sum of ΔH_f of gaseous atoms

$\Delta H_2 = -\Delta H_{atom}$

$\Delta H_3 = \Delta H$ for conversion of gas to compound in its standard state

elements ⟶ compound
(standard state) ΔH_f° (standard state)

If the standard state of the compound is the gas, ΔH_3 can be ignored.

Example 11.2

Calculate the value of ΔH_f° for dimethyl ether vapor,

$$
\begin{array}{ccc}
\text{H} & & \text{H} \\
| & & | \\
\text{H-C-O-C-H} \\
| & & | \\
\text{H} & & \text{H}
\end{array}
$$

Solution

The reaction whose ΔH we wish to calculate is as follows, with the alternative path indicated below it.

$$2C(s, \text{ graphite}) + 3H_2(g) + \tfrac{1}{2}O_2(g) \xrightarrow{\Delta H_f^\circ} C_2H_6O(g)$$

with lower-path steps:

$2\Delta H_f[C(g)] \qquad 6\Delta H_f[H(g)] \qquad \Delta H_f[O(g)] \qquad -\Delta H_{atom}$

$$2C(g) \quad + \quad 6H(g) \quad + \quad O(g) \longrightarrow$$

The minus sign appears before the ΔH_{atom} because in the direction of the arrow the process is the reverse of atomization. When a reaction is reversed, remember that the sign of ΔH is reversed, too.

The sum of all ΔH's along the lower path must equal the ΔH along the upper path (that is, ΔH_f°).

$$\Delta H_f^\circ = 2\Delta H_f[C(g)] + 6\Delta H_f[H(g)] + \Delta H_f[O(g)] - \Delta H_{atom}$$

From Table 10.2 we can get the first three terms on the right.

$$\Delta H_f^\circ = 2(+715 \text{ kJ}) + 6(+218 \text{ kJ}) + (+249 \text{ kJ}) - \Delta H_{atom}$$

ΔH_{atom} is calculated from the number and kind of bonds in the molecule.

6(C—H) bonds	6(415) =	2490 kJ
2(C—O) bonds	2(356) =	712 kJ
	ΔH_{atom} =	3202 kJ

Substituting,

$$\Delta H_f^\circ = 1430 \text{ kJ} + 1308 \text{ kJ} + 249 \text{ kJ} - 3202 \text{ kJ}$$

$$\Delta H_f^\circ = -215 \text{ kJ}$$

Since we are dealing with the formation of one mole of product,

$$\Delta H_f^\circ = -215 \text{ kJ/mol}$$

(The experimentally measured value is -185 kJ/mol.)

Self-Test

13. Calculate the atomization energy for acetic acid,

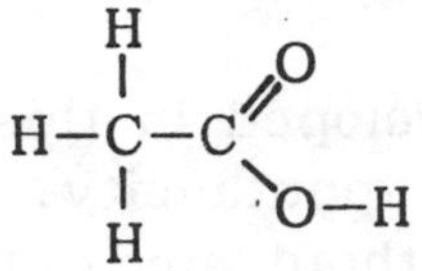

$$
\begin{array}{c}
\quad\;\; H \\
\quad\;\; | \\
H-C-C \overset{\displaystyle O}{\underset{\displaystyle O-H}{\diagup\diagdown}} \\
\quad\;\; | \\
\quad\;\; H
\end{array}
$$

_____

14. Use the data in Tables 11.2 and 11.3 to calculate ΔH°_{f} for $C_2H_6(g)$ in kJ/mol. Compare your value with that found in Table 11.1. The structure of C_2H_6 is

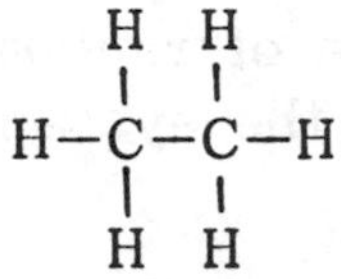

$$
\begin{array}{c}
H \; H \\
| \;\; | \\
H-C-C-H \\
| \;\; | \\
H \; H
\end{array}
$$

_____

15. Use the data in Tables 11.2 and 11.3 to calculate ΔH°_{f} for $C_6H_6(\ell)$. The heat of vaporization of C_6H_6 is 8.19 kcal/mol. The molecule is represented as a resonance hybrid.

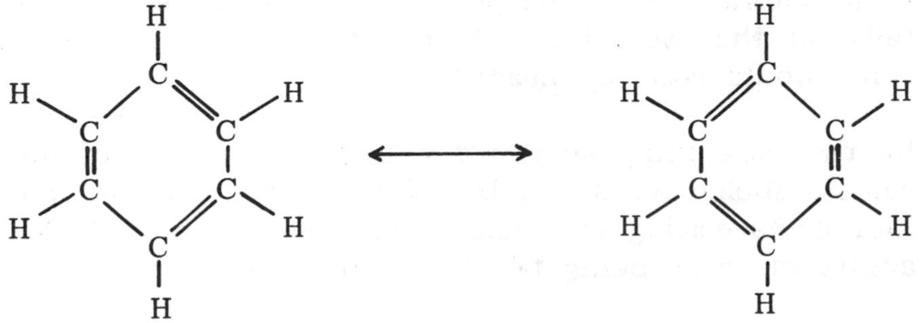

How does your calculated value compare with that found in Table 11.1?

_____

New Terms

atomization energy

11.7   SPONTANEITY OF CHEMICAL REACTIONS

Objectives

To examine the factors that favor spontaneity in chemical

and physical changes.

Review

The main point developed in this section is that there are two factors that influence spontaneity. A change tends to occur spontaneously (that is, without any outside help) if it takes place with a lowering of its energy. Therefore, exothermic reactions tend to occur spontaneously. It is also found that a physical or chemical change is favored when the system proceeds from a less probable state to one of greater probability. This corresponds to an increase in the degree of randomness, or disorder, in the arrangement of particles in the system.

Self-Test

16. You know that in tossing a coin there is an equal probability of it coming up either heads or tails. Suppose you had two coins, labeled A and B, and were to toss them. Make a table showing all of the possible combinations of heads and tails for the two coins. What is the probability that both coins would come up heads?

   _____

17. Do the same thing as you did in Question 16, but for 4 coins, labeled A, B, C, D. What is the probability of all four coins coming up heads when tossed? What is the probability of there being two heads and two tails?

   _____

New Terms

spontaneous change

11.8   ENTROPY

Objectives

To describe a thermodynamic state function that can be associated with randomness in a chemical or physical system.

Review

The entropy, S, is a state function that is a measure of the probability of occurrence of a given state of a system. Because the existence of a disordered or random arrangement of particles is more probable than a highly ordered arrangement, entropy is said to increase as the disorder of the system increases.

The entropy change, ΔS, is given by

$$\Delta S = q_{rev}/T$$

where q_{rev} is the heat absorbed by a system if it were to go from one state to another by a reversible process and T is the absolute temperature at which the heat is added to the system.

New Terms

entropy

11.9   THE SECOND LAW OF THERMODYNAMICS

Objectives

To obtain an equation that incorporates the two factors that favor spontaneous changes (i.e., decrease in energy, increase in entropy).

Review

Chemical and physical changes are favored when accompanied by a decrease in energy. A change also tends to be spontaneous if it occurs with an increase in entropy.

The second law of thermodynamics states, in essence, that whenever a spontaneous change occurs, the entropy of the universe increases.

The Gibbs free energy is defined as

$$G = H - TS$$

At constant temperature and pressure,

$$\Delta G = \Delta H - T\Delta S$$

Remember that for a process to be spontaneous ΔG must be negative. In terms of the right side of the above equation, the difference, $\Delta H - T\Delta S$, must be negative. This section considers three possibilities:

(1) ΔH negative, ΔS positive: ΔG will be negative at all temperatures. The process will be spontaneous regardless of the temperature.

(2) ΔH positive, ΔS negative: ΔG will be positive at all temperatures. The process cannot be spontaneous at any temperature.

(3) ΔH and ΔS both positive or both negative: the sign of ΔG depends on the value of T. If ΔH and ΔS are both positive, for example, ΔG becomes the difference between two positive quantities (ΔH and $T\Delta S$). At high temperature $T\Delta S$ can be larger than ΔH; ΔG will be negative and the change will be spontaneous. At low temperature ΔH will be larger than $T\Delta S$; ΔG will be positive and the change won't be spontaneous.

Self-Test

18.   What will be the sign of ΔG at high temperature if both ΔH and ΔS are negative?

_____

19.   The conversion of liquid CCl_4 to gaseous CCl_4 at 1 atm occurs with $\Delta H = +32.8$ kJ/mol and $\Delta S = +95.0$ J/mol K. Above what temperature should bubbles be able to form within the liquid spontaneously?

_____

New Terms

Gibbs free energy
second law of thermodynamics

11.10   FREE ENERGY AND USEFUL WORK

Objectives

To see how ΔG is related to the useful work that can be extracted from a system during a spontaneous change.

Review

      The free energy change, ΔG, is equal to the <u>maximum</u> work that can be obtained from a spontaneous change. Remember that this maximum work can only be obtained if the change is reversible. All real systems undergo changes in an irreversible manner; therefore, the amount of work that can be extracted from a real system is somewhat less than the maximum predicted by ΔG.

New Terms

11.11   STANDARD ENTROPIES AND FREE ENERGIES

Objectives

      To establish standard entropies and free energies of substances. You should learn the third law of thermodynamics. You should also learn to calculate ΔG° for reactions from ΔG_f°.

Review

      The third law of thermodynamics states that for any pure substance, $S = 0$ at 0 K. Absolute values of entropy can be obtained and some are given in Table 11.4. Values of ΔG° can be gotten from calculated ΔH° and ΔS°, the latter obtained by suitably combining the S° of products and reactants.

      ΔS° = (Sum S° products) - (Sum S° reactants)

      Table 11.5 tabulates values of ΔG_f°. These can be used in Hess's law type calculations to compute values of ΔG° for reactions.

      ΔG° = (Sum of ΔG_f° products) - (Sum of ΔG_f° reactants)

Self-Test

20. Calculate values of ΔS_f° for the following in units of J/mol K.

    (a) $Al_2O_3(s)$ _____

(b) $C_3H_8(g)$ _____

(c) $PbSO_4(s)$ _____

21. The standard entropy of $CS_2(g)$ is 237.7 J/mol K. Use the data in Tables 11.1 and 11.4 to calculate ΔG_f° of $CS_2(g)$.

_____

22. Calculate ΔG° in kilojoules for the following reactions:

(a) $H_2SO_4(\ell) + CaO(s) \longrightarrow CaSO_4(s) + H_2O(\ell)$ _____

(b) $Ag(s) + 2HNO_3(\ell) \longrightarrow AgNO_3(s) + NO_2(g) + H_2O(\ell)$

_____

New Terms

third law of thermodynamics
standard entropies
standard free energies of formation

11.12   FREE ENERGY AND EQUILIBRIUM

Objectives

To see the relationship between free energy and equilibrium. You should learn the qualitative relationship between ΔG° and the position of equilibrium.

Review

The main points in this section are that $\Delta G = 0$ when a system is at equilibrium, and that ΔG° is related to the position of equilibrium. Review the discussion centering on Figures 11.13 and 11.14. Notice that even with a ΔG° that is positive, some reaction occurs. This is because ΔG starts out negative heading in the direction of the products. Also note that the distinction is made here between ΔG° and ΔG. ΔG° is the difference between the free energy of the products in their standard states and the free energy of the reactants in their standard states. We are using ΔG, on the other hand, to stand for the way that the free energy is changing as we move along the free energy curve. Our earlier discussion about spontaneity applies in the sense that the free energy must be decreasing (ΔG negative) when the process is occurring spontaneously. Once the minimum is reached, equilibrium is established because the system cannot climb out of the free energy well.

In Figure 11.14, notice that the position of equilibrium is determined by the sign (and magnitude) of ΔG°. Only when ΔG° is negative do we observe the formation of significant amounts of products. In this way we can use ΔG° as a predictor of reaction spontaneity.

Self-Test

23.   Sketch free energy curves for the reactions:

(a) $N_2(g) + \frac{1}{2}O_2(g) \longrightarrow N_2O(g)$ $\Delta G° = +104$ kJ

(b) $H_2(g) + Cl_2(g) \longrightarrow 2HCl(g)$ $\Delta G° = -191$ kJ

24.   Is the reaction, $N_2(g) + O_2(g) \longrightarrow 2NO(g)$, "spontaneous" at room temperature? Is it more or less "spontaneous" at 1000°C? (Assume $\Delta H°$ and $\Delta S°$ are essentially independent of temperature.)

New Terms

Answers to Self-Test Questions

1.(a) isothermal (b) state function (c) adiabatic (d) specific heat (e) heat capacity (f) molar heat capacity 2. 464 J (111 cal) 3.(a) 64 liter-atm (b) 80 liter-atm 4. 59.9 Calories
5.(a) 802 kJ/mol (b) 802 kJ/mol (PΔV = 0) 6. -1284 kJ/mol
7. -69.8 kcal 8. -1777 kJ 9. -177 kJ 10.(a) 429.1 kJ (-102.6 kcal) (b) -30 kJ (-6.6 kcal) (c) -451 kJ (-107.8 kcal)
11. -59.4 kcal/mol 12. N_2O has a positive $\Delta H°_f$, so when it serves as a reactant it tends to make ΔH more negative (since ΔH of reactants are subtracted from those of the products).
13. 3136 kJ/mol (749.6 kcal/mol) 14. Calc. $\Delta H°_f$ = -100 kJ/mol
15. Calc. $\Delta H°_f$ = +209 kJ/mol (+49.7 kcal/mol); actual $\Delta H°_f$ = +49.0 kJ/mol

16.

A	B
H	H
H	T
T	H
T	T

Probability of both heads = 1/4 = 0.25

17.

A	B	C	D	
H	H	H	H	*
H	T	H	H	
H	H	T	H	
H	H	H	T	
H	T	T	H	**
H	T	H	T	**
H	H	T	T	**
H	T	T	T	
T	H	H	H	
T	T	H	H	**
T	H	T	H	**
T	H	H	T	**
T	T	T	H	
T	T	H	T	
T	H	T	T	
T	T	T	T	

16 possible combinations

* = all heads
** = 2 heads, 2 tails

probability all heads = 1/16 = 0.0625

probability 2 heads, 2 tails = 6/16 = 0.375

18. ΔG will be positive   19. 345 K (72°C).   When $\Delta G = 0$, $\Delta H = T\Delta S$, $T = \Delta H/\Delta S$.   Above this T, ΔG will be negative.

20.(a) -313.1 J/mol K   (b) -269.6 J/mol K   (c) -358 J/mol K

21. $\Delta G_f^0 = \Delta H_f^0 - (298\ K)\Delta S_f^0$; $\Delta G_f^0 = +66.8$ kJ/mol   22.(a) -262 kJ
(b) -57.3 kJ

23.(a)   (b)

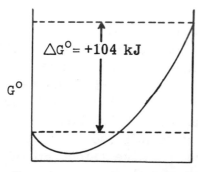

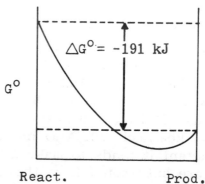

24. Since $\Delta G° = +174$ kJ, the reaction is not "spontaneous" because very little NO will be produced. At 1000°C (1273 K),

$$\Delta G = \Delta H° - (1273 \text{ K})\Delta S°$$
$$= +181 \text{ kJ} - (1273 \text{ K})(+24.7 \text{ J/mol K})$$
$$= +181 \text{ kJ} - 31.4 \text{ kJ} = +150 \text{ kJ}$$

Since ΔG is smaller, the reaction will be a little more "spontaneous" at the higher temperature.

12 CHEMICAL KINETICS

This chapter concerns itself with the rates at which chemical reactions take place. This subject is important for several reasons. First, unless a reaction proceeds at a measurable rate, no products will be observed regardless of how thermodynamically favorable the reaction might be. Second, a study of reaction rates and the factors that influence them provides insight into the sequence of chemical steps that occurs to produce the overall net reaction. Third, a study of the effect of temperature on reaction rate gives information about energy changes that occur along the path from reactants to products.

Remember the four factors that influence the rate of a reaction.
1. The chemical nature of the reactants and products.
2. The concentration of the reactants.
3. The effect of temperature.
4. The influence of catalysts.

12.1 REACTION RATES AND THEIR MEASUREMENT

Objectives

To see what is meant by reaction rate and to see how it is measured. You should also learn the units used to express reaction rates.

Review

The term "reaction rate" describes how fast the concentrations of reactants or products change with time and is usually expressed in the units mol/liter second (mol liter^{-1} s^{-1}). For most reactions the rate changes (decreases) as the reactants are consumed.

The rate of reaction can be obtained from a graph of the concentration of a reactant (or product) versus time (see Figure 12.2). The rate at some time, t, is obtained from the slope of the tangent to the concentration-time curve at time t. The procedure is described in Figure 12.2.

Remember that square brackets, [ ], denote molar concentration.

Self-Test

1. The rate of a reaction was found to be 3.0 x 10^{-4} mol/liter second. What would be the rate if it were expressed in the units, mol/liter minute?

    _____

2. From the concentration vs time curve below, estimate the rate of reaction at t = 10 s.

    _____

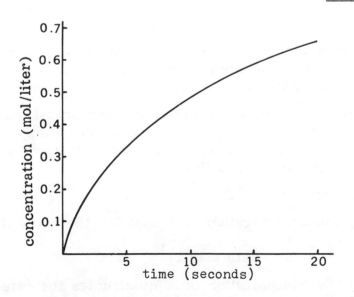

3. The sequence of chemical reactions that provides the net overall change in a chemical reaction is called

    _____

4. List the four factors that affect the rate of a chemical reaction.

    _____

    _____

New Terms

mechanism
reaction rate

12.2   RATE LAWS

Objectives

To learn what is meant by the term, rate law, and to see how it can be obtained from experimental data.

Review

The rate law for a reaction such as

$$A + B \longrightarrow products$$

is

$$Rate = k[A]^x[B]^y$$

where k is the rate constant. The exponents, x and y, are called the order of the reaction with respect to A and B, respectively. The overall order is (x + y).

It is very important to remember that the actual values of x and y can only be obtained by experiment. These experiments involve observing what effect altering the concentrations of the reactants has on the rate of reaction.

For a reaction, $A \longrightarrow B$,

(1) if doubling the concentration of A doubles the rate, the exponent is 1.
$$Rate = k[A]^1$$

(2) if doubling the concentration of A quadruples the rate, the exponent is 2.
$$Rate = k[A]^2$$

Review Examples 12.1 and 12.2 in the text to see how these rules are applied. Notice that once the exponents in the rate law have been established, the rate constant can be evaluated from any of the sets of data.

Self-Test

5. Below are some typical rate laws. What are the orders of the reactions and the orders with respect to each reactant?

(a) Rate = $k[NO]^2[Br_2]$

$2NO + Br_2 \longrightarrow 2NOBr$ _____

(b) Rate = $k[NO]^2[H_2]$

$2NO + 2H_2 \longrightarrow N_2 + 2H_2O$ _____

6. The following data were collected for the reaction,

$$A + 2B \longrightarrow C + D$$

| concentrations | | rate |
A	B	(mol/liter s)
0.10	0.10	2.0×10^{-4}
0.10	0.20	4.0×10^{-4}
0.20	0.20	1.6×10^{-3}
0.30	0.20	3.6×10^{-3}

(a) The rate law for the reaction is _____

(b) The value of the rate constant is _____

(c) The units of the rate constant are _____

New Terms

order of reaction   rate constant
rate law

12.3   CONCENTRATION AND TIME: HALF-LIVES

Objectives

To learn how the concentrations of the reactants are related to time for first-order and second-order reactions. You should learn the concept of half-life and know how it is computed for first- and second-order reactions.

Review

For a first-order reaction with the rate law, rate = k[A], the concentration at any time, t, after the start of the reaction (t = 0) is

$$\ln \frac{[A]_o}{[A]_t} = kt$$

or, in terms of common logs,

$$2.303 \log \frac{[A]_o}{[A]_t} = kt$$

For a second-order reaction with a rate law, rate = k[B]^2, the relationship between concentration and time is

$$\frac{1}{[B]_t} - \frac{1}{[B]_o} = kt$$

The half-life, $t_{\frac{1}{2}}$, of a given reactant is the time it takes for the reactant's concentration to be reduced to half of its initial value. For a first-order reaction, remember that $t_{\frac{1}{2}}$ is independent of the reactant's initial concentration, and the first and successive half-lives are all equal. They can be computed from the rate constant

$$t_{\frac{1}{2}} = \frac{\ln 2}{k} = \frac{0.693}{k}$$

The half-life in a second-order reaction does depend on the reactant's initial concentration. Doubling the initial concentration halves the $t_{\frac{1}{2}}$; halving the initial concentration doubles the $t_{\frac{1}{2}}$. This means that successive half-lives in a second-order reaction increase by a factor of 2. In other words the second half-life is double the first because its initial concentration is only half the first. Similarly, the third half-life is double the second, and so on. The half-life can be calculated by the expression

$$t_{\frac{1}{2}} = \frac{1}{k[B]_o}$$

Self-Test

7. In a certain first-order reaction having the rate law, rate = k[C], the rate constant has a value of 3.00×10^{-4} s^{-1}. If the initial concentration of C is 0.50 mol/liter,

(a) what will be its concentration after 30 minutes?

_____

(b) what is $t_{\frac{1}{2}}$ for the reaction?   _____

(c) what will be the concentration of C after 4 half-lives?

_____

8. In a certain second-order reaction having the rate law, rate = k[D]^2, the reaction was begun with a concentration of D equal to 1.00 M. The rate constant for the reaction is $k = 5.0 \times 10^{-2}$ liter mol^{-1} s^{-1}.

(a) What is the initial half-life of D?   _____

(b) How long will it take for the concentration of D to be reduced to 0.125 M?

_____

New Terms

half-life

12.4   COLLISION THEORY

Objectives

　　To obtain a theory that accounts quantitatively for the dependence of reaction rate on concentration.

Review

　　The basis for the collision theory is the notion that molecules must collide in order to react with one another. We can predict the rate law if we know what collisions take place during a reaction. Usually reactions occur in a series of steps before the ultimate products have been formed and we don't actually know what these steps are. In fact, one of the goals of kinetics is to give us a way of guessing intelligently at what these steps might be.

Remember, <u>if</u> we have the following collision processes, their rate laws are:

$$A + B \longrightarrow products \qquad\qquad Rate = k[A][B]$$

$$A + A \longrightarrow products$$

or $\qquad 2A \longrightarrow products \qquad\qquad Rate = k[A]^2$

<u>New Terms</u>

collision theory
bimolecular collision

12.5   REACTION MECHANISMS

<u>Objectives</u>

To see how collision theory helps us choose between alternative possible mechanisms for a reaction. You should learn the meaning of "rate-determining step."

<u>Review</u>

The individual reactions that make up a mechanism are called elementary processes. The slowest step in the mechanism is called the rate-determining step because the final products can't be formed any faster than the products of the slowest step.

Obtaining a satisfactory mechanism for a reaction is a very difficult task. A chemist must draw on all his experience and knowledge to arrive at a set of elementary processes that both make sense chemically and fit the experimentally determined rate law. With your limited chemical background you can't be expected to derive chemically reasonable mechanisms. However, within reason, you should be able to decide from a comparison of predicted and experimentally found rate laws whether a given mechanism is possible. You should also be able to decide which step in a mechanism must be the slow step in order to yield the correct rate law.

Remember that the predicted rate law should only include the reactants in the overall equation. Any intermediate products should not appear. In the very simple mechanisms that we are dealing with you can get the exponents in the predicted rate law by adding together all of the steps up to and including the rate-determining step. The coefficients of the reactants at this point are the exponents in the predicted rate law. Try this with the mechanism proposed for the reaction between NO and H_2 at the beginning of the section in the text. The second reaction is the rate-determining step.

Self-Test

9. Consider the following mechanisms:

$$(1)\quad 2A \longrightarrow Q$$
$$(2)\quad Q + B \longrightarrow C + D$$
$$(3)\quad B + D \longrightarrow 2M$$

(a) What is the equation for the overall reaction?

_____

(b) What is the rate law if step 1 is rate-determining?

_____

(c) What is the rate law if step 2 is rate-determining?

_____

10. The following mechanism was proposed to account for a chemical reaction. The rate law was found experimentally to be: Rate = $k[R]^2$

$$2R \longrightarrow X + Y$$
$$X + Z \longrightarrow T + U$$
$$U + Z \longrightarrow P$$

Which is the rate-determining step? _____

New Terms

elementary process
rate-determining step

12.6   EFFECTIVE COLLISIONS

Objectives

To understand why reaction rates are nearly always much less than the rate of collision between molecules.

Review

The first thing to keep in mind here is that except for a very few cases, reactions proceed at a much slower rate than we would at first expect on the basis of the frequency of molecular collisions. Two factors tend to limit the number of collisions that are effective. One is the energy of the colliding molecules; the other is related to the orientation of the molecules when they collide.

New Terms

12.7   TRANSITION STATE THEORY

Objectives

To follow the energy changes that take place during an effective collision. Also, to see how the rate of reaction is affected by the energy required to produce an effective collision. You should become familiar with the potential energy diagram for a reaction.

Review

The minimum kinetic energy required between two colliding molecules in order to produce an effective collision is called the activation energy, E_a. Review the energy diagrams in Figures 12.6 to 12.8. You should be able to identify:

(1) the potential energy of the reactants
(2) the potential energy of the products
(3) the activation energy for the forward reaction

(4) the activation energy for the reverse reaction
(5) the heat of reaction
(6) whether the forward reaction is endo- or exothermic

   The species that exists at the top of the potential energy diagram (i.e., the high-energy species that is formed during an effective collision) is called the activated complex. The peak on the potential energy diagram is called the transition state. Transition state theory is concerned with the characteristics (geometry, energy) of the activated complex.

   As a general rule, reactions having high activation energies tend to occur slowly, whereas fast reactions usually have low activation energies.

Self-Test

11. Identify the following by the numbers on the potential energy diagram which follows.

   (a) _____ the potential energy of the products

   (b) _____ the activation energy for the forward reaction

   (c) _____ the heat of reaction

   (d) _____ the potential energy of the reactants

   (e) _____ the activation energy of the reverse reaction

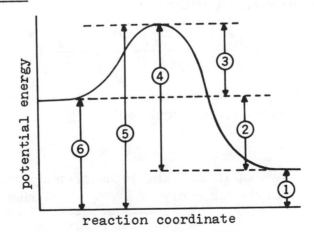

New Terms

activation energy
reaction coordinate

activated complex
transition state

12.8   EFFECT OF TEMPERATURE ON REACTION RATE

Objectives

To learn how and why temperature affects the rate of
reaction. You should be able to calculate the activation
energy from rate constants at two different temperatures.

Review

In very nearly every case, increasing the temperature increases the rate of reaction. This is because molecules move
faster, on the average, at higher temperatures and more molecular collisions have the minimum kinetic energy to produce a net
chemical change.

The Arrhenius equation is

$$k = Ae^{-E_a/RT}$$

From this can be derived the expression,

$$\ln \frac{k_1}{k_2} = \frac{E_a}{R} \left(\frac{1}{T_2} - \frac{1}{T_1} \right)$$

or, in terms of common logarithms,

$$\log \frac{k_1}{k_2} = \frac{E_a}{2.303\ R} \left(\frac{1}{T_2} - \frac{1}{T_1} \right)$$

If rate constants k_1 and k_2 are known at temperatures T_1 and T_2,
the activation energy can be calculated.

Self-Test

12. As a rough rule of thumb, the rates of many reactions approximately double for every _____ °C rise in temperature.

13. The rate constant of a reaction at 15°C is 1.3×10^{-5} liter/
mol s, while at 50°C its rate constant is 8.0×10^{-3} liter/mol
s. What is the E_a for the reaction? _____

14. The activation energy for a certain reaction was found to be 25.0 kcal/mol. At 25°C the rate constant is 2×10^{-3} s^{-1}. What is the rate constant at 50°C?

_____

New Terms

Arrhenius equation

12.9   CATALYSTS

Objectives

To define what a catalyst is and to understand how it functions in affecting the rate of reaction.

Review

A catalyst is a substance that alters the rate of a reaction by providing an alternative path (mechanism) from reactants to products that has a lower activation energy than the uncatalyzed reaction. It is important to remember that a catalyst functions by changing the mechanism.

A homogeneous catalyst exists in the same phase as the reactants. Biological enzymes are examples of homogeneous catalysts that promote specific biochemical reactions.

A heterogeneous catalyst exists as a separate phase from the reactants and products and appears to function by adsorbing reactant molecules on its surface where the reaction can somehow proceed more readily. Heterogeneous catalysts are widely used in industrial applications because they don't have to be separated later from the products of reaction as would homogeneous catalysts.

An inhibitor is a substance that becomes adsorbed on the surface of the catalyst and thereby decreases, or inhibits, its activity. The catalyst is then said to be poisoned.

Self-Test

15. How does a catalyst alter the activation energy of a reaction?

_____

16. What type of catalyst is present in a catalytic muffler?

_____

17. What type of catalyst is a biological enzyme?

_____

New Terms

catalyst   heterogeneous catalyst
homogeneous catalyst   inhibitor

12.10 CHAIN REACTIONS

Objectives

To examine a type of mechanism that often occurs in
systems with very complicated rate laws. You should
learn that these reactions tend to propagate themselves
once they have been started. You should learn the
kinds of steps responsible for initiation, continuation,
and termination of the chain.

Review

Chain reactions often involve free radicals - molecules or
ions that contain unpaired electrons. These unpaired electrons
tend to pair up with electrons in other molecules or ions, and
free radicals are very reactive. The kinds of reactions that may
be found in a chain mechanism are:

(1) initiation - this is the step that generates the first free
radical.

(2) propagation - a product is formed plus another free radical.
This continues the chain.

(3) inhibition - this is a step that slows down the rate of formation of products. In the mechanism in the text it removes
some product but still generates a free radical so the chain
can continue.

(4) termination - this is a step that removes free radicals and
therefore interrupts the chain.

Self-Test

18. The reaction of methane with chlorine ($CH_4 + Cl_2 \longrightarrow$ $CH_3Cl + HCl$) is believed to occur by the following mechanisms. Identify the nature of each of these reactions.

(a) $Cl_2 \xrightarrow{\text{light}} 2Cl\cdot$ _____

(b) $Cl\cdot + CH_4 \longrightarrow HCl + CH_3\cdot$ _____

(c) $CH_3\cdot + Cl_2 \longrightarrow CH_3Cl + Cl\cdot$ _____

(d) $Cl\cdot + Cl\cdot \longrightarrow Cl_2$ _____

(e) $CH_3\cdot + CH_3\cdot \longrightarrow C_2H_6$ _____

(f) $CH_3\cdot + Cl\cdot \longrightarrow CH_3Cl$ _____

New Terms

free radical   propagation step
chain reaction   termination step
initiation step   inhibition step

Answers to Self-Test Questions

1. 1.8×10^{-2} mol/liter minute 2. 0.026 mol/liter s 3. the mechanism 4. nature of the reactants, concentration of reactants, temperature, catalysts 5.(a) second order in NO, first order in Br_2, third order overall (b) second order in NO, first order in H_2, third order overall 6.(a) Rate = $k[A]^2[B]$ (b) k = 0.2 (c) liter^2/mol^2 s   7.(a) 0.29 mol/liter (b) 2310 s = 38.5 min (c) 0.031 mol/liter 8.(a) 20 seconds (b) 140 seconds 9.(a) $2A + 2B \longrightarrow C + 2M$ (b) Rate = $k[A]^2$ (c) Rate = $k[A]^2[B]$ 10. The first step 11.(a) 1 (b) 3 (c) 2 (d) 6 (e) 4 12. 10°C 13. 34.0 kcal/mol (143 kJ/mol) 14. 5.2×10^{-2} s^{-1} 15. a catalyst provides a different, low energy mechanism 16. heterogeneous catalyst 17. homogeneous catalyst 18.(a) initiation (b) propagation (c) propagation (d) termination (e) termination (f) termination

13 CHEMICAL EQUILIBRIUM

As a chemical reaction proceeds, the concentrations of the reactants decrease and the rate of the forward reaction decreases. At the same time, the concentrations of the products increase and the rate of the reverse reaction increases. Eventually both reactions occur at the same rate and dynamic equilibrium is achieved.

There is a simple relationship between the concentrations of reactants and products in an equilibrium system. This chapter deals with that relationship, how it can be understood from the standpoint of thermodynamics, and how it can be used in calculations relating equilibrium concentrations.

13.1 THE LAW OF MASS ACTION

Objectives

To establish the relationships between reactant and product concentrations in a chemical equilibrium. You should note that this section views the equilibrium law as a purely experimentally measurable phenomenon without attempting to present an explanation for it.

Review

Remember that the mass action expression (or reaction quotient) for any chemical reaction can be written from the balanced equation. The concentrations of the products always appear in the numerator, raised to powers that are equal to their coefficients in the balanced equation. The concentrations of reactants are multiplied together in the denominator where they are each raised to powers equal to their coefficients in the balanced equation.

At equilibrium, at a given temperature, the mass action expression for a given reaction is always equal to the same number, the equilibrium constant. Remember, there are no restrictions on the individual equilibrium concentrations. The only requirement is that when they are substituted into the mass action expression, the resulting fraction must equal the equilibrium constant. This defines the equilibrium law for the reaction.

For gaseous reactions the equilibrium constant expression can be written in terms of concentrations, thereby giving K_c, or in terms of partial pressures, giving K_p.

Self-Test

1. Write the mass action law (equilibrium law) giving both K_c and K_p for each of the following gaseous reactions:

    (a) $PCl_5 \rightleftharpoons PCl_3 + Cl_2$

    (b) $Br_2 + SO_2 + 2H_2O \rightleftharpoons H_2SO_4 + 2HBr$

    (c) $6XeF_4 + 12H_2O \rightleftharpoons 2XeO_3 + 4Xe + O_2 + 24HF$

    (d) $2NO_2 + F_2 \rightleftharpoons 2NO_2F$

2. Below are equilibrium concentrations of NO_2 and N_2O_4. The equilibrium equation is: $N_2O_4(g) \rightleftharpoons 2NO_2(g)$

Experiment	$[N_2O_4]$	$[NO_2]$
1	4.46×10^{-2} M	3.11 M
2	1.50×10^{-3} M	0.571 M
3	2.30×10^{-7} M	7.06×10^{-3} M
4	1 M	14.7 M

Show that they obey the mass action law. What is the value
of the equilibrium constant?

_____

New Terms

law of mass action   reaction quotient
mass action expression   equilibrium law
equilibrium constant

13.2 THE EQUILIBRIUM CONSTANT

Objectives

To see how the magnitude of K provides an immediate
qualitative estimate of the extent to which a reaction
proceeds toward completion.

Review

When K is large the reaction proceeds far toward completion; when K is small hardly any products are present at equilibrium.

Self-Test

4. Arrange the following reactions in order of increasing tend-
    ency to proceed toward completion:

    (a) $CO(g) + Cl_2(g) \rightleftharpoons COCl_2(g)$ $K = 5 \times 10^9$

    (b) $N_2O_4(g) \rightleftharpoons 2NO_2(g)$ $K = 217$

    (c) $2SO_2(g) + O_2(g) \rightleftharpoons 2SO_3(g)$ $K = 8 \times 10^{25}$

    (d) $2HCl(g) \rightleftharpoons H_2(g) + Cl_2(g)$ $K = 3.1 \times 10^{-17}$

    _____

New Terms

13.3   THERMODYNAMICS AND CHEMICAL EQUILIBRIUM

Objectives

To relate, quantitatively, the standard free energy change for a reaction to the equilibrium constant. After completing this section you should be able to compute K from $\Delta G°$, and vice versa.

Review

The free energy change for a reaction is related to the mass action expression (Q) by the equation,

$$\Delta G = \Delta G° + RT \ln Q \qquad \text{(13.3 in text)}$$

or, in terms of common logarithms,

$$\Delta G = \Delta G° + 2.303\ RT \log Q$$

Important equations to remember are the following:

For gaseous reactions,

$$\Delta G° = -RT \ln K_p \qquad \text{(13.5 in text)}$$

or

$$\Delta G° = -2.303\ RT \log K_p$$

For reactions in solution,

$$\Delta G° = -RT \ln K_c \qquad \text{(13.6 in text)}$$

or

$$\Delta G° = -2.303\ RT \log K_c$$

The K's that you compute using these equations are called thermodynamic equilibrium constants. In using the equations, you must be sure to use the correct value of R, according to the energy units of $\Delta G°$.

$$R = 1.987\ \text{cal mol}^{-1}\ \text{K}^{-1}$$
$$R = 8.314\ \text{J mol}^{-1}\ \text{K}^{-1}$$

Also, for $\Delta G°$ be sure to convert kJ to J or kcal to cal, and be sure to express the temperature in kelvins. This is emphasized again in the following examples.

Example 13.1

What is the thermodynamic equilibrium constant, K_p, for the reaction, $2HBr(g) + Cl_2(g) \rightleftharpoons 2HCl(g) + Br_2(g)$, at 25°C? Work the problem using $\Delta G°$ in kilocalories.

Solution

First we calculate $\Delta G°$ from the appropriate $\Delta G_f°$ in Table 11.5.

$$\Delta G° = 2\Delta G_f°(HCl) - 2\Delta G_f°(HBr)$$

$$\Delta G° = 2 \text{ mol}(-22.8 \text{ kcal/mol}) - 2 \text{ mol}(-12.7 \text{ kcal/mol})$$

$$\Delta G° = -20.2 \text{ kcal}$$

Using a scientific calculator, it is simplest to work with natural logarithms. We therefore solve Equation 13.5 for $\ln K_p$.

$$\ln K_p = \frac{-\Delta G°}{2.303 \text{ RT}}$$

In these calculations, remember:

(1)   use $R = 1.987$ cal mol^{-1} K^{-1} if $\Delta G°$ is in kcal; use $R = 8.314$ J/mol K if $\Delta G°$ is in kJ.

(2)   make sure you express $\Delta G°$ in calories (or J), not kcal (or kJ).

(3)   use the absolute temperature (298 K in this question).

Substituting,

$$\ln K_p = \frac{-(-20200)}{(1.987)(298)} = 34.1$$

To take the antilog, we use the e^x function on the calculator.

$$K_p = e^{34.1}$$
$$= 6 \times 10^{14}$$

If you don't have a scientific calculator with logarithm functions, you must use Equation 13.6 in the text. In that case,

$$\log K_p = \frac{-(-20200)}{2.303(1.987)(298)} = 14.8$$

Taking the antilogarithm gives

$$K_p = 6 \times 10^{14}$$

(Logarithms and antilogarithms are discussed in Appendix A of the textbook.)

Self-Test

5. Calculate the value of the thermodynamic equilibrium constant at 25°C for the reaction, $H_2(g) + I_2(g) \rightleftharpoons 2HI(g)$

6. What is the value of $\Delta G°$ if at 25°C a reaction has an equilibrium constant equal to 1.0?

7. What is the value of $\Delta G°$ in both kJ and kcal if at 25°C a reaction has an equilibrium constant of 1.5×10^{-12}?

New Terms

thermodynamic equilibrium constant

13.4   THE RELATIONSHIP BETWEEN K_P and K_c

Objectives

To see how we can convert from K_p to K_c, and vice versa.

Review

Remember the relationship,

$$K_p = K_c(RT)^{\Delta n_g}$$

where Δn_g is the change in the number of moles of <u>gas</u> on going from reactants to products in the balanced equation. Review the sample calculation in Example 13.4 in the text. Notice that in this calculation $R = 0.0821$ liter atm mol^{-1} K^{-1}. This is to make units cancel properly.

Self-Test

8. The reaction, $N_2(g) + 3H_2(g) \rightleftharpoons 2NH_3(g)$, has $K_p = 7.2 \times 10^5$ at 25°C. What is the value of K_c? _____

9. The reaction, $H_2(g) + I_2(g) \rightleftharpoons 2HI(g)$, has $K_p = 0.35$ at 25°C. What is the value of K_c? _____

New Terms

13.5   HETEROGENOUS EQUILIBRIA

Objectives

   To see how the mass action expression can be simplified in cases of equilibrium between two or more pure phases.

Review

   Remember that the concentrations of pure liquid or solid phases are not included in the mass action expression. This is because they are constant and are included in the equilibrium constant.

Self-Test

10. Write the equilibrium constant expression for K_c for the following:

   (a) $2H_2(g) + O_2(g) \rightleftharpoons 2H_2O(\ell)$ _____

   (b) $CO_2(g) + Li_2CO_3(s) + H_2O(g) \rightleftharpoons 2LiHCO_3(s)$

   _____

   (c) $Cl_2(g) + 2KBr(s) \rightleftharpoons 2KCl(s) + Br_2(g)$

   _____

New Terms

heterogeneous equilibrium

13.6   LE CHÂTELIER'S PRINCIPLE AND CHEMICAL EQUILIBRIUM

Objectives

To learn how to apply Le Châtelier's principle to changes in concentration, pressure, and temperature in chemical systems.

Review

(1)   When a reactant or product is added to a system at equilibrium, the position of equilibrium shifts toward the opposite side of the equation.

(2)   Decreasing the concentration of a reactant or product causes the position of equilibrium to shift in the direction of the substance removed.

(3)   Increasing the pressure by decreasing the volume shifts the position of equilibrium in the direction of the fewest number of moles of gas.

(4)   An increase in temperature causes the position of equilibrium to shift in the direction of the endothermic reaction.

A very important point to remember is that the only thing that changes K for a reaction is a change in temperature!

Two final observations are made in this section. One is that adding an inert (unreactive) gas to a system, without changing the volume, has no effect on the position of equilibrium. The second is that a catalyst has no effect on the position of equilibrium. It only increases the speed with which the system reaches equilibrium.

Self-Test

11.   Use the letters, I = increase, D = decrease, N = no change, to indicate what effect each of the following changes will have upon the concentration of $SO_2(g)$ in the system,

$$2SO_2(g) + O_2(g) \rightleftharpoons 2SO_3(g) \qquad \Delta H = -193 \text{ kJ}$$

(a)   adding $O_2(g)$ _____

(b)   adding $SO_3(g)$ _____

(c) removing $SO_3(g)$ _____

(d) increasing the temperature   _____

(e) increasing the volume of the container   _____

(f) adding helium   _____

(g) adding a catalyst   _____

12. Which, if any, of the changes described in Question 11 will alter the equilibrium constant?

_____

New Terms

13.7 EQUILIBRIUM CALCULATIONS

Objectives

To learn how to use the equilibrium constant expression in numerical calculations.

Review

There are basically two kinds of calculations that you have to learn to do. One is to calculate K from either equilibrium concentrations or information from which you can deduce equilibrium concentrations. The other is to calculate information about equilibrium concentrations, having at your disposal the value of K. There are six sample calculations in the text illustrating these types of problems (Examples 13.6 to 13.11). These are worked out in great detail. Two additional sample calculations follow. A very important thing to notice in all of these is that the concentrations (or algebraic quantities representing concentrations) that are substituted into the mass action expression in the equilibrium law are always equilibrium concentrations.

The first example deals with the calculation of K from a set of equilibrium concentrations.

Example 13.2

At a certain temperature, 0.15 mol CO(g) and 0.15 mol H_2O(g) are introduced into a 2.0-liter container. At equilibrium the CO(g) concentration was measured to be 0.042 mol/liter. What is the value of K_c for the reaction,

$$CO(g) + H_2O(g) \rightleftharpoons H_2(g) + CO_2(g)$$

Solution

First let's write the equilibrium constant expression.

$$K_c = \frac{[H_2][CO_2]}{[CO][H_2O]}$$

We need equilibrium concentrations of each gas. First let's calculate the initial concentrations. Remember, always work with concentrations!

$$[CO] = 0.15 \text{ mol}/2.0 \text{ liter} = 0.075 \text{ M}$$
$$[H_2O] = 0.15 \text{ mol}/2.0 \text{ liter} = 0.075 \text{ M}$$

How about the products? Their initial concentrations were zero. At equilibrium they are present because some CO and H_2O reacted to produce them. How much reacted?

We are given the equilibrium concentration of CO, 0.042 M. This is less than we started with. The difference between what is present at equilibrium and what we started with is the amount that reacted.

$$\text{Amount of CO reacted} = 0.075 \text{ mol/liter} - 0.042 \text{ mol/liter}$$
$$= 0.033 \text{ mol/liter}$$

How much H_2O reacted with the CO? From the equation we can see that the answer must also be 0.033 mol/liter, and the amount of H_2O remaining at equilibrium must be

$$[H_2O]_{equilibrium} = 0.075 \text{ M} - 0.033 \text{ M} = 0.042 \text{ M}$$

For the reactants at equilibrium,

$$[H_2O] = 0.042 \text{ M}$$
$$[CO] = 0.042 \text{ M}$$

How about the products? When 0.033 mol/liter of CO reacts, it must produce 0.033 mol/liter of both H_2 and CO_2. This we can see from the coefficients in the balanced equation. At equilibrium, then,

$$[H_2] = 0.033 \text{ M}$$

$$[CO_2] = 0.033 \text{ M}$$

The reasoning that we've just gone through is simplified somewhat by setting up a concentration table similar to that in Example 13.6 in the text. Using the values in this problem,

	Initial Concentration	Change	Equilibrium Concentrations
CO	0.075 M	-0.033 M	0.042 M
H_2O	0.075 M	-0.033 M	0.042 M
H_2	0	+0.033 M	0.033 M
CO_2	0	+0.033 M	0.033 M

Substituting these equilibrium concentrations into the mass action expression, we can calculate K.

$$K_c = \frac{(0.033)(0.033)}{(0.042)(0.042)}$$

$$K_c = 0.62$$

The second example asks you to calculate equilibrium concentrations using the known value of K. This kind of question involves some very simple algebra. Don't panic over it. If you take your time and think it through slowly, you should be able to learn how to approach this kind of problem. Don't try to memorize how to solve specific problems. If you do, it only takes a small change in the problem to trip you up.

Example 13.3

In the last example we found that at a particular temperature $K_c = 0.62$ for the reaction,

$$CO(g) + H_2O(g) \rightleftharpoons H_2(g) + CO_2(g)$$

The equilibrium concentrations were: $[CO] = [H_2O] = 0.042$ M and $[H_2] = [CO_2] = 0.033$ M. Suppose an additional 0.010 mol/

liter of CO(g) and 0.010 mol/liter of $H_2O(g)$ are intoduced into the container. What will the new equilibrium concentrations become?

Solution

As soon as the additional CO and H_2O are added, the equilibrium is upset, and we can treat the new concentrations as initial concentrations.

initial concentrations

$[CO]$ $= 0.042 + 0.010 = 0.052$ M
$[H_2O] = 0.042 + 0.010 = 0.052$ M
$[H_2]$ $= 0.033$ M
$[CO_2] = 0.033$ M

We now take into account that a reaction will take place that brings the system back to equilibrium. Let's let x equal the number of moles/liter of CO that will react. The CO concentrations at equilibrium will therefore have been diminished by x. The H_2O concentration will also have been decreased by x mol/ liter (from the stoichiometry of the balanced equation). Similarly, the H_2 and CO_2 concentrations will each increase by x.

equilibrium concentrations

$[CO]$ $= 0.052 - x$
$[H_2O] = 0.052 - x$
$[H_2]$ $= 0.033 + x$
$[CO_2] = 0.033 + x$

The reasoning that we have just gone through is the key to solving the problem. It is made easier with the concentration table.

	Initial Concentrations	Change	Equilibrium Concentrations
CO	0.052 M	-x	(0.052 - x) M
H_2O	0.052 M	-x	(0.052 - x) M
H_2	0.033 M	+x	(0.033 + x) M
CO_2	0.033 M	+x	(0.033 + x) M

Next, we substitute the equilibrium quantities into the mass action expression (see the last example),

$$0.62 = \frac{(0.033 + x)(0.033 + x)}{(0.052 - x)(0.052 - x)} = \frac{(0.033 + x)^2}{(0.052 - x)^2}$$

The simplest way to solve this problem is to take the square root of both sides of the equation; $\sqrt{0.62} = 0.79$

$$0.79 = \frac{0.033 + x}{0.052 - x}$$

Now multiply both sides by $(0.052 - x)$.

$$0.79(0.052 - x) = 0.033 + x$$
$$0.041 - 0.79\,x = 0.033 + x$$
$$0.041 - 0.033 = x + 0.79x$$
$$0.008 = 1.79\,x$$
$$0.004 = x$$

The equilibrium concentrations become

$$[CO] = [H_2O] = 0.052 - 0.004 = 0.048 \text{ M}$$
$$[H_2] = [CO_2] = 0.033 + 0.004 = 0.037 \text{ M}$$

When K for a reaction is very small or very large, the position of equilibrium lies very close to either the reactants or products. Recognizing this fact can sometimes help us simplify the algebra involved in an equilibrium problem, as shown in Example 13.11.

Self-Test

13. At approximately 1700°C the equilibrium concentrations of the reactants and product in the equation,
$$N_2(g) + O_2(g) \rightleftharpoons 2NO(g)$$
are $[N_2] = 1.0 \times 10^{-4}$ M, $[O_2] = 2.5 \times 10^{-5}$ M
$[NO] = 7.1 \times 10^{-7}$ M
What is K_c for this reaction?   _____

14. In a furnace operating at 1700°C the concentrations of N_2 and O_2 are 4.0×10^{-4} M and 1.0×10^{-5} M, respectively. At this temperature $K_c = 2.0 \times 10^{-4}$ (the answer to the last question!). What concentration of NO(g) will be present in the gases escaping from the furnace? (Assume the N_2 and O_2 concentrations are equilibrium concentrations.)

_____

15. At a certain temperature $K_c = 0.5$ for the reaction,

$$H_2(g) + I_2(g) \rightleftharpoons 2HI(g)$$

If 0.40 mol of H_2 and 0.40 mol of I_2 are placed into a 1.0-liter container at this temperature, what will be the equilibrium concentration of each gas?

_____

16. At 25°C, the reaction, $2HCl(g) \rightleftharpoons H_2(g) + Cl_2(g)$ has $K_c = 3.2 \times 10^{-34}$. If 0.400 mol of HCl is placed in a 4.00-liter container at this temperature, what will be the concentrations of H_2 and Cl_2 after equilibrium has been reached?

_____   _____

New Terms

Answers to Self-Test Questions

1.(a)

$$K_c = \frac{[PCl_3][Cl_2]}{[PCl_5]} \qquad K_p = \frac{p_{PCl_3}\, p_{Cl_2}}{p_{PCl_5}}$$

(b)

$$K_c = \frac{[H_2SO_4][HBr]^2}{[Br_2][SO_2][H_2O]^2} \qquad K_p = \frac{p_{H_2SO_4}\, p_{HBr}^2}{p_{Br_2}\, p_{SO_2}\, p_{H_2O}^2}$$

(c)

$$K_c = \frac{[XeO_3]^2[Xe]^4[O_2][HF]^{24}}{[XeF_4]^6[H_2O]^{12}}$$

$$K_p = \frac{p_{XeO_3}^2\, p_{Xe}^4\, p_{O_2}\, p_{HF}^{24}}{p_{XeF_4}^6\, p_{H_2O}^{12}}$$

(d)

$$K_c = \frac{[NO_2F]^2}{[NO_2]^2[F_2]} \qquad K_p = \frac{p_{NO_2F}^2}{p_{NO_2}^2 \, p_{F_2}}$$

2. K = 217 when values are substituted in the expression,

$$K = \frac{[NO_2]^2}{[N_2O_4]}$$

3. Equilibrium constants can be given for reactions without having to also specify the form of the mass action expression.

4. d < b < a < c   5. K = 0.35   6. zero, since log 1.0 = 0

7. +16.1 kcal (+67.5 kJ)   8. $\Delta n_g = -2$, $K_c = 4.3 \times 10^8$

9. $\Delta n_g = 0$, $K_c = K_p = 0.35$

10.(a)

$$K_c = \frac{1}{[H_2]^2[O_2]} \qquad \text{(b)} \quad K_c = \frac{1}{[CO_2][H_2O]} \qquad \text{(c)} \quad K_c = \frac{[Br_2]}{[Cl_2]}$$

11.(a) D (b) I (c) D (d) I (e) I (f) N (g) N

12. only (d) will change K

13. $K_c = 2.0 \times 10^{-4}$

14. 8.9×10^{-7} M

15. $[H_2] = [I_2] = 0.30$ M; $[HI] = 0.20$ M

16. $[H_2] = [Cl_2] = 1.8 \times 10^{-18}$ M.

14 ACIDS AND BASES

The acid-base concept is one of the most useful ways of correlating a large amount of what otherwise might appear to be unrelated chemical information. In this chapter the acid-base definition presented in Chapter 6 is expanded to cover acid-base reactions in solvents other than water, and even to reactions in the absence of any solvent whatsoever. Before reading the first section, review the three properties that acids and bases have in common, which are given in the introduction to the chapter in the text.

14.1   THE ARRHENIUS DEFINITION OF ACIDS AND BASES

Objectives

> To review the definition of acids and bases in the solvent, water. You should learn what characterizes an acid and a base in water, and you should learn the neutralization reaction that occurs in aqueous solutions.

Review

Remember, in water an acid produces H_3O^+ and a base produces OH^-. Many acids contain hydrogen; for example, HCl, H_2SO_4, HNO_3. Other substances react with water to produce acids; these are called acid anhydrides. Bases often contain OH^- (e.g., NaOH, $Ba(OH)_2$). Some molecular substances react with water to generate OH^-. An example is NH_3. In general,

nonmetal oxides are acid anhydrides
metal oxides are basic anhydrides

Self-Test

1. Write the chemical equation for the reaction of the following with water:

    (a) HBr   _____

    (b) H_2SO_4 _____

    _____

    (c) N_2H_4 _____

    (d) O^{2-} _____

2. Identify the following as acid (A) or basic (B) anhydrides.

    (a) CaO   _____   (d) Li_2O _____

    (b) SO_2 _____   (e) N_2O_3 _____

    (c) CO_2 _____   (f) P_4O_6 _____

3. Write chemical equations showing the reactions of the following with water.

    (a) SO_3 _____

    (b) P_4O_{10} _____

    (c) BaO   _____

    (d) N_2O_3 _____

4. What is the net neutralization reaction between an acid and a base in aqueous solution?

    _____

5. What three general properties do acids and bases exhibit?

    (a) _____

    (b) _____

    (c) _____

New Terms

Arrhenius concept of acids and bases
acid anhydride
basic anhydride

14.2   BRØNSTED-LOWRY DEFINITION OF ACIDS AND BASES

Objectives

To define acids and bases in terms of the transfer of a
proton. You should learn to identify acid-base conjugate
pairs in an acid-base reaction. You should learn what an
amphiprotic (amphoteric) substance is.

Review

Learn the Brønsted-Lowry definitions of acid and base:

acid - proton (H^+) donor
base - proton acceptor

Acid-base reactions can be looked upon as reversible.

$$\text{Acid (X)} + \text{Base (Y)} \rightleftharpoons \text{Base (X)} + \text{Acid (Y)}$$

$$HX + Y \rightleftharpoons X^- + HY^+$$

There are two conjugate acid-base pairs in this reaction.

Acid (X) - Base (X)   (HX, X^-)
Base (Y) - Acid (Y)   (Y, HY^+)

Let's look at a concrete example,

$$HNO_3 + H_2O \rightleftharpoons H_3O^+ + NO_3^-$$

acid   base   acid   base

conjugate pair

conjugate pair

Notice that the only difference between the members of a conjugate pair is a single proton. All other atoms are identical. Also,
note that the acid has one more hydrogen than the base.

Example 14.1

From the list below, choose those two which form a conjugate acid-base pair. Which one is the acid?

$$H_2SO_4,\ OH^-,\ HSO_3^-,\ SO_3^{2-},\ SO_4^{2-}$$

Solution

The only two species that differ from each other by only one hydrogen are

$$HSO_3^- \text{ and } SO_3^{2-}$$

The one with the most hydrogen (HSO_3^-) is the acid.

An amphiprotic (amphoteric) substance can act as either an acid or a base. The most common example is water.

acid	base	
H_3O^+	H_2O	OH^-
	acid	base

Review the autoionization reactions on Page 450 of the text. Also, note that the Brønsted-Lowry definition permits acid-base reactions in the absence of a solvent. The reaction between HCl(g) and NH_3(g) is an example.

Some metal ions, especially those with a high charge, produce acidic solutions. An example is the Al^{3+} ion, which exists as $Al(H_2O)_6^{3+}$ in solution.

Self-Test

6. In each of the following indicate whether the underlined substance is behaving as an acid (A) or a base (B).

   (a) $H_2O + \underline{HC_2H_3O_2} \rightleftharpoons H_3O^+ + C_2H_3O_2^-$ ______

   (b) $\underline{CN^-} + HCl \rightleftharpoons HCN + Cl^-$ ______

   (c) $\underline{NH_3} + O^{2-} \rightleftharpoons NH_2^- + OH^-$ ______

7. For each part in Question 6, identify both acid-base conjugate pairs. Underline the acid in each of them.

(a) _____, _____

(b) _____, _____

(c) _____, _____

8. Which of the following <u>could not</u> serve as amphiprotic substances?
    (a) NH_3 (b) CN^- (c) O^{2-} (d) HSO_4^- (e) SO_4^{2-}

    _____

9. Write a chemical equation to show why solutions containing $Cr(H_2O)_6^{3+}$ are slightly acidic.

    _____

New Terms

Brønsted acid

Brønsted base

conjugate acid

conjugate base

conjugate acid-base pair

amphiprotic

amphoteric

autoionization reaction

14.3   STRENGTHS OF ACIDS AND BASES

Objectives

To compare the relative strengths of acids and bases.

Review

The position of equilibrium in the reaction,

$$HA + B^- \rightleftharpoons HB + A^-$$

allows us to compare the relative strengths of the acids and bases. A strong acid tends to give up its proton more readily than a weak acid. If HA is stronger than HB, the position of equilibrium lies to the right. Similarly, a strong base is able to capture protons more readily than a weak base. If B^- is a stronger base than A^-, the B^- captures more protons than the A^- and the position of equilibrium again lies to the right. In general the position of equilibrium lies in the direction of the weaker acid and base.

$$\text{strong acid} + \text{strong base} \rightleftharpoons \text{weak acid} + \text{weak base}$$

The leveling effect occurs when comparing the strengths of strong acids in a basic solvent, or strong bases in an acidic solvent. Any solvent that makes it possible to distinguish between the strengths of acids or bases is called a differentiating solvent.

Self-Test

10. The following are equations for the reaction of some Brønsted acids with water. The equilibrium constants are also given. If necessary, review the relationship between position of equilibrium and the magnitude of K as given in Section 13.2 (see Page 228 in the Study Guide and Page 427 in the text). Arrange the acids in order of increasing acid strength.

   (a) $HNO_2 + H_2O \rightleftharpoons H_3O^+ + NO_2^-$ $K = 4.5 \times 10^{-4}$

   (b) $HF + H_2O \rightleftharpoons H_3O^+ + F^-$ $K = 6.5 \times 10^{-4}$

   (c) $HCN + H_2O \rightleftharpoons H_3O^+ + CN^-$ $K = 4.9 \times 10^{-10}$

   (d) $H_3PO_4 + H_2O \rightleftharpoons H_3O^+ + H_2PO_4^-$ $K = 7.5 \times 10^{-3}$

   (e) $HC_3H_5O_2 + H_2O \rightleftharpoons H_3O^+ + C_3H_5O_2^-$ $K = 1.4 \times 10^{-5}$

11. The ions, Br^- and Cl^-, are very weak Brønsted bases. What property would a solvent have to have to be considered a differentiating solvent for these ions?

New Terms

leveling effect   differentiating solvent
leveling solvent

14.4 LEWIS ACIDS AND BASES

Objectives

   To provide a still more general definition of an acid and

a base. You should learn how the Lewis definition can be used to explain acid-base reactions.

Review

Under the Lewis definition we have:

base - electron pair donor
acid - electron pair acceptor

A Lewis acid and base react with each other by the formation of a coordinate covalent bond. Lewis bases tend to be species that have lone pairs of electrons and completed octets. Lewis acids tend to be substances that can be considered "electron deficient," or which can make themselves electron deficient by rearrangement of their electrons. The latter is illustrated by SO_3 when it reacts with oxide ion.

Lewis acid-base reactions are often viewed as displacement reactions.

(1) nucleophilic displacement - one base displaces another.

acid
H
[:S:]2−
+
O
H
[:S - H]−
+
[:O - H]−
incoming
base
(nucleophile)
outgoing
base

$$S^{2-} + H_2O \longrightarrow HS^- + OH^-$$

(2) electrophilic displacement - one acid displaces another.

:Cl:
Cl5Sb
+
N:
:O:
[Cl5Sb:Cl:]−
+
[:N=O:]+
incoming
acid
(electrophile)
outgoing
acid

$$SbCl_5 + NOCl \longrightarrow SbCl_6^- + NO^+$$

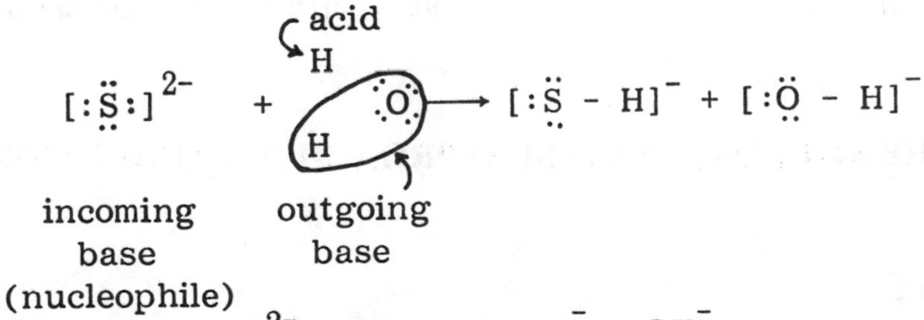

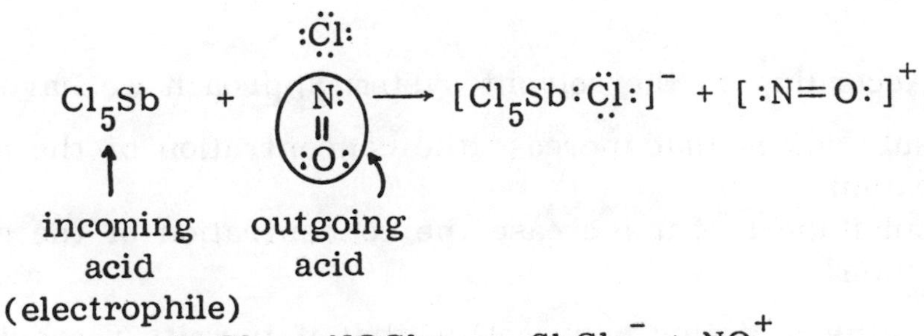

Self-Test

12. Use electron-dot formulas to show how the reaction of O^{2-} with SO_2 to produce SO_3^{2-} can be considered the reaction between a Lewis acid and base.

13. Diethyl ether (shown below) reacts with boron trichloride to form an addition compound. Use Lewis structures to show how this is a Lewis acid-base reaction.

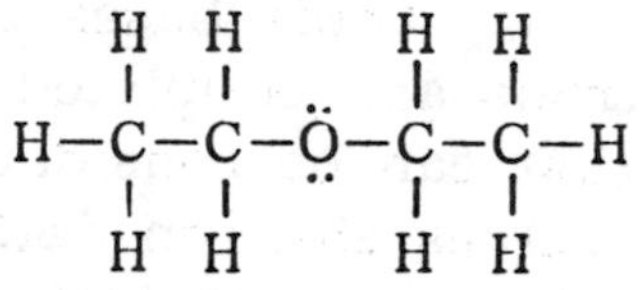

$$
\begin{array}{ccccc}
& H & H & H & H \\
& | & | & .. & | & | \\
H- & C- & C- & \ddot{O}- & C- & C-H \\
& | & | & .. & | & | \\
& H & H & H & H \\
\end{array}
$$

diethyl ether

New Terms

Lewis acid	nucleophilic displacement
Lewis base	electrophile
nucleophile	electrophilic displacement

14.5   THE SOLVENT SYSTEM APPROACH TO ACIDS AND BASES

Objectives

To define acids and bases in terms of the ions produced by the autoionization of the solvent.

Review

According to the solvent system approach we have:

acid - substances that increase the concentration of the solvent cation.

base - substances that increase the concentration of the solvent anion.

The usefulness of this approach is that it permits reasoning by analogy from one solvent to another. Compare, for example, the neutralization reactions in water and ammonia discussed on Page 458.

Self-Test

14. Bromine trifluoride, BF_3, undergoes autoionization,

$$BF_3 + BF_3 \rightleftharpoons BF_2^+ + BF_4^-$$

There are also the reactions,

$$SbF_5 + BF_3 \longrightarrow SbF_6^- + BF_2^+$$
$$KF + BF_3 \longrightarrow K^+ + BF_4^-$$

How would you classify (acid or base) the following?

(a) SbF_5 _____   (b) KF _____

15. In liquid ammonia, what is the reaction between NH_4Cl and KNH_2? _____

16. Sodium oxide is the basic anhydride of NaOH in aqueous solutions. What is the basic anhydride of $NaNH_2$ in liquid ammonia? _____

17. In some situations, liquid sulfur dioxide behaves as if it undergoes autoionization,

$$SO_2 + SO_2 \rightleftharpoons SO^{2+} + SO_3^{2-}$$

How would you expect the following to behave in liquid SO_2?

(a) Na_2SO_3 _____   (b) $SOCl_2$ _____

(c) What reaction, if any, would occur between Na_2SO_3 and $SOCl_2$ in liquid SO_2? _____

New Terms

solvent system

14.6 SUMMARY

Objectives

To bring into perspective the features of the different acid-base definitions discussed in this chapter.

New Terms

Answers to Self-Test Questions

1.(a) $HBr + H_2O \longrightarrow H_3O^+ + Br^-$

 (b) $H_2SO_4 + H_2O \longrightarrow H_3O^+ + HSO_4^-$

   $HSO_4^- + H_2O \longrightarrow H_3O^+ + SO_4^{2-}$

 (c) $N_2H_4 + H_2O \longrightarrow N_2H_5^+ + OH^-$

 (d) $O^{2-} + H_2O \longrightarrow 2\,OH^-$

2.(a) B (b) A (c) A (d) B (e) A (f) A

3.(a) $SO_3 + H_2O \longrightarrow H_2SO_4$

 (b) $P_4O_{10} + 6H_2O \longrightarrow 4H_3PO_4$

 (c) $BaO + H_2O \longrightarrow Ba(OH)_2$

 (d) $N_2O_3 + H_2O \longrightarrow 2HNO_2$

4. $H_3O^+ + OH^- \longrightarrow 2H_2O$, or simply, $H^+ + OH^- \longrightarrow H_2O$

5.(a) neutralization (b) reaction with indicators (c) catalysis

6.(a) acid (b) base (c) acid

7.(a) H_2O, H_3O^+; $HC_2H_3O_2$, $C_2H_3O_2^-$

 (b) CN^-, $\overline{HCN}$; $\overline{HCl, Cl^-}$

 (c) NH_3, $\overline{NH_2^-}$; $\overline{O^{2-}}$, $\overline{OH^-}$

8. b, c and e. They can't be Brønsted acids because they have no hydrogen.

9. $Cr(H_2O)_6^{3+} + H_2O \rightleftharpoons Cr(H_2O)_5OH^{2+} + H_3O^+$

10. $HCN < HC_3H_5O_2 < HNO_2 < HF < H_3PO_4$

11. It would have to be a very acidic solvent so that some protonation of each would occur. The weaker of the two would tend to be protonated the least.

12.

sulfite ion

13.

$$\underset{\underset{\substack{H-C-H \\ | \\ H-C-H \\ | \\ H}}{\overset{\displaystyle H \;\; H}{\underset{\displaystyle H \;\; H}{H-C-C-\ddot{O}:}}}{} \;\; + \;\; \underset{\overset{\displaystyle Cl}{\underset{\displaystyle Cl}{B-Cl}}}{} \;\; \longrightarrow \;\; \underset{\underset{\substack{H-C-H \\ | \\ H-C-H \\ | \\ H}}{\overset{\displaystyle H \;\; H \;\;\;\; Cl}{\underset{\displaystyle H \;\; H \;\; | \;\; Cl}{H-C-C-\ddot{O}-B-Cl}}}}{}$$

14. (a) acid - it increases BF_2^+ (b) base - it increases BF_4^-

15. $NH_4Cl + KNH_2 \longrightarrow KCl + 2NH_3$

16. Na_3N or Na_2NH

17. (a) base   (b) acid, $SOCl_2 \longrightarrow SO^{2+} + 2Cl^-$

(c) These react together, $Na_2SO_3 + SOCl_2 \longrightarrow 2SO_2 + 2NaCl$

This is a neutralization reaction in liquid SO_2.

15 ACID-BASE EQUILIBRIA IN AQUEOUS SOLUTION

This chapter deals quantitatively with the equilibria involving the autoionization of water and the dissociation of weak acids and bases. These are important in any aqueous system, particularly biological ones where many important molecules behave as weak acids or bases. Many of the numerical problems in this chapter and the next require the application of some simple algebra. Don't panic! Follow the procedures shown in the worked-out examples which explain how to approach these problems. The important thing is to try not to rush - don't skip steps in the reasoning. If you proceed slowly, you should be able to master the material in Chapters 15 and 16.

15.1 IONIZATION OF WATER, pH

Objectives

To establish the quantitative criteria for equilibrium in the autoionization of water and to devise a system for expressing small concentrations of H_3O^+.

Review

In this section you saw that the equilibrium condition for the ionization of water reduces to

$$K_w = [H^+][OH^-] = 1.0 \times 10^{-14}$$

In any solution in which water is the solvent this condition holds. You are expected to know the value, $K_w = 1.0 \times 10^{-14}$.

In pure water, $[H^+] = [OH^-] = 1.0 \times 10^{-7}$ M. When H^+ or OH^- are present from another source (e.g., an acid or a base in the solution), $[H^+] \neq [OH^-]$. The product of concentrations, however, must equal K_w.

In general, for any quantity X,

$$pX = -\log X$$

For the hydrogen ion concentration,

$$pH = -\log [H^+]$$

Similarly, for hydroxide ion,

$$pOH = -\log [OH^-]$$

Remember that the sum of pH and pOH equals pK_w.

$$pH + pOH = pK_w = 14.0$$

You should be able to calculate pH given $[H^+]$, and also $[H^+]$ given pH. Review Examples 15.2, 15.3 and 15.4 in the text. To solve these problems it is necessary to refer to the table of logarithms (Appendix B) or use an electronic calculator having a LOG function. Actually, the availability of electronic calculators has greatly simplified pH calculations. If you have a calculator with a LOG key, the pH can be found by entering the $[H^+]$ and depressing the LOG key. The pH, preceded by a minus sign, will appear on the display. To obtain the $[H^+]$ from pH you can use the Y^X key, because $[H^+] = 10^{-pH}$. Enter 10 as the value of Y, depress the Y^X key, then enter the negative of the pH (i.e., -pH) as the value of X. Finally, depress the "=" key and the $[H^+]$ will appear on the display. The actual sequence of operations, of course, may differ somewhat for different brands of calculators. Also, note that common logarithms are used to compute pH, not natural logarithms!

Self-Test

1. Calculate the $[OH^-]$ in the following solutions:

    (a) $[H^+] = 1.0 \times 10^{-8}$ M _____

    (b) $[H^+] = 4.2 \times 10^{-12}$ M _____ _____

(c) $[H^+] = 8.4 \times 10^{-2}$ M _____

2. Calculate the $[H^+]$ in the following solutions:

(a) $[OH^-] = 1.0 \times 10^{-4}$ M _____

(b) $[OH^-] = 2.8 \times 10^{-9}$ M _____

(c) $[OH^-] = 6.7 \times 10^{-12}$ M _____

3. Calculate the pH and pOH of solutions with the following concentrations:

(a) $[H^+] = 2.0 \times 10^{-6}$ M _____

(b) $[H^+] = 8.5 \times 10^{-9}$ M _____

(c) $[OH^-] = 3.4 \times 10^{-6}$ M _____

4. Calculate the $[H^+]$ in the following solutions:

(a) pH = 8.50 _____

(b) pH = 13.34 _____

(c) pOH = 13.34 _____

New Terms

ion product constant   pH
ionization constant   pOH
dissociation constant pK_w

15.2 DISSOCIATION OF WEAK ELECTROLYTES

Objectives

To deal quantitatively with equilibria involving the dissociation of weak electrolytes, in particular weak acids and bases. You should be able to calculate K, given equilibrium concentrations. You should be able to calculate equilibrium concentrations from K and the concentration of the weak electrolyte.

Review

For a weak acid that undergoes the ionization,

$$HA \rightleftharpoons H^+ + A^-$$

we have

$$K_a = \frac{[H^+][A^-]}{[HA]}$$

For a weak base, B, that reacts with the solvent,

$$B + H_2O \rightleftharpoons HB^+ + OH^-$$

we have

$$K_b = \frac{[HB^+][OH^-]}{[B]}$$

Examples 15.5 and 15.6 in the text show you how to calculate the equilibrium constant if you have information that allows you to compute equilibrium concentrations.

To calculate equilibrium concentrations from K you must know the amount of weak acid or base placed in the solution. The following example is typical.

Example 15.1

What are the concentrations of H^+, OCl^- and HOCl in a solution labeled 0.10 M HOCl? For HOCl, $K_a = 3.1 \times 10^{-8}$.

Solution

First, write the equation for the equilibrium.

$$HOCl \rightleftharpoons H^+ + OCl^-$$

The next step for problems of this type is to set up our table of concentrations as shown in the example problems in the text.

	initial concentration	change	equilibrium concentration
H^+	0.0	+x	x
OCl^-	0.0	+x	x
HOCl	0.10	−x	0.10 − x

| | This column contains the concentrations of solutes placed into the solution. | This column represents the changes that occur because of reaction. | This column gives the equilibrium concentrations. |

The entries in the table are obtained by the following reasoning. If no dissociation occurred, we would have [HOCl] = 0.10 M. HOCl does dissociate, however, and we want to calculate how much. The approach, then, is to let x equal the number of moles per liter of HOCl that have dissociated when equilibrium is reached. The concentrations at equilibrium, then, would be

$$[HOCl] = 0.10 - x \qquad \text{(note that the concentration of HOCl is diminished by the quantity that is lost upon dissociation)}$$

No H^+ or OCl^- were placed in the solution; they are there at equilibrium because of the dissociation of the HOCl.

$$[OCl^-] = x$$
$$[H^+] = x$$

(note that $[OCl^-] = [H^+]$ because they are formed in a 1:1 ratio. The amounts produced per liter are equal to the amount per liter of HOCl that has dissociated)

The equilibrium constant expression is

$$K_a = \frac{[H^+][OCl^-]}{[HOCl]} = 3.1 \times 10^{-8}$$

Substituting our expressions for the equilibrium concentrations from the last column of the table gives

$$\frac{(x)(x)}{(0.10 - x)} = 3.1 \times 10^{-8}$$

Expanding this out will give a quadratic equation. The problem can be simplified however. From the magnitude of K_a we know that very little HOCl will dissociate. What we do is assume that the amount that dissociates is negligible; that is,

$$0.10 - x \approx 0.10$$

This simplification gives

$$\frac{(x)(x)}{(0.10)} = 3.1 \times 10^{-8}$$

$$x^2 = (0.10)(3.1 \times 10^{-8}) = 3.1 \times 10^{-9}$$

$$x^2 = 31 \times 10^{-10}$$

$$x = \sqrt{31} \times 10^{-5}$$

$$x = 5.6 \times 10^{-5}$$

(Note that to take the square root without an electronic calculator, you must have the exponent on 10 divisible by 2.)

Finally, we have the equilibrium concentrations,

$$[HOCl] = 0.10 - 5.6 \times 10^{-5} = 0.10 \text{ M}$$

$$[H^+] = [OCl^-] = 5.6 \times 10^{-5} \text{ M}$$

Observe that x is indeed negligible compared to 0.10 when the difference is rounded to the proper number of significant figures. This justifies our assumption.

Sometimes you will have to deal with problems involving solutions containing a weak acid or base plus a salt of that acid or base. An example is $HC_2H_3O_2$ and $NaC_2H_3O_2$. In working problems of this type you should remember the following:

(1) Salts are completely dissociated. The concentration of the ion produced by the salt is entered in the "initial concentration" column.

(2) In the problems that you will encounter only one of the ions produced by the salt is important. The other is a spectator ion.

Review Examples 15.8 and 15.9 in the text.

Self-Test

5. A 0.20 M solution of propionic acid, $HC_3H_5O_2$, has a hydrogen ion concentration of 1.7×10^{-3} M. What is the value of K_a for propionic acid?

6. A 1.0 molar solution of the weak base methyl amine, CH_3NH_2, has a pH of 12.32. What is the value of K_b?

7. Aniline, $C_6H_5NH_2$, has an ionization constant, $K_b = 3.8 \times 10^{-10}$. Aniline reacts with water according to the equation, $C_6H_5NH_2 + H_2O \rightleftharpoons C_6H_5NH_3^+ + OH^-$. What are the OH^- and H^+ concentrations in a 0.01 M solution of $C_6H_5NH_2$?

8. Calculate the percent dissociation in a 0.25 M solution of HCN. $K_a = 4.9 \times 10^{-10}$

    _____

9. What is the value of pK_a for HCN? K_a is given in Question 8.

    _____

10. What is the pH of a 0.20 M solution of lactic acid, $HC_3H_5O_3$, that also contains 0.30 M $NaC_3H_5O_3$? For lactic acid, $K_a = 1.38 \times 10^{-4}$.

    _____

New Terms

acid dissociation constant
acid ionization constant

base ionization constant
percent dissociation

15.3 DISSOCIATION OF POLYPROTIC ACIDS

Objectives

To consider equilibria for acids that dissociate in two or more steps. You should learn how to calculate the hydrogen ion concentration as well as the concentrations of other ions produced in the equilibria.

Review

For a diprotic acid there are two equilibria and two equilibrium constants. For H_2S, for example,

$$H_2S \rightleftharpoons H^+ + HS^- \qquad\qquad K_{a1} = \frac{[H^+][HS^-]}{[H_2S]}$$

$$HS^- \rightleftharpoons H^+ + S^{2-} \qquad\qquad K_{a2} = \frac{[H^+][S^{2-}]}{[HS^-]}$$

Remember, to calculate the hydrogen ion concentration in a solution of a polyprotic acid you always use K_{a1}. This is because the second and succeeding ionization steps almost always produce much less H^+ than the first (i.e., $K_{a1} \gg K_{a2} \gg K_{a3}\ldots$). You should also use K_{a1} to calculate the concentration of the anion produced in the first step.

For diprotic acids you should use K_{a2} to calculate the

concentration of the anion produced in the second dissociation (e.g., S^{2-} in the second ionization of H_2S). If the diprotic acid is the only solute in the solution, the concentration of the anion, A^{2-} (e.g., S^{2-}) is equal to K_{a2}.

In a solution of a diprotic acid, H_2A, a combined expression

$$K_{a1}K_{a2} = \frac{[H^+]^2[A^{2-}]}{[H_2A]}$$

can be used <u>provided that</u> the values of any two of the three concentrations that appear in the expression are known. These are the <u>only</u> conditions under which this combined expression can be applied.

<u>Self-Test</u>

11. Calculate the concentrations of the ions, H^+, HA^- and A^{2-}, in a 0.10 M solution of the weak acid, H_2A. $K_{a1} = 2.0 \times 10^{-4}$, $K_{a2} = 5.0 \times 10^{-9}$

_____

12. In a solution of H_2SO_3 the sulfite concentration is 0.20 M and the pH is 6.50. What is the concentration of H_2SO_3 in the solution?

_____

<u>New Terms</u>

15.4 BUFFERS

<u>Objectives</u>

To learn how the pH of a solution can be controlled by a mixture of a weak acid or base and its salt. You should learn how to calculate the relative amounts of acid (or base) and salt needed to give a desired pH. You should also be able to calculate the effect on pH produced by addition of small amounts of strong acid or base to a buffer.

Review

Remember, a buffer is a mixture of a weak acid and a weak base. This can be obtained by mixing a weak acid with one of its salts, since the anion of the acid is a weak base. A buffer can also be made by mixing a weak base with one of its salts. The H^+ concentration in an acid buffer can be calculated from the equilibrium constant expression.

$$[H^+] = K_a \frac{[HA]}{[A^-]}$$

From this the Henderson-Hasselbalch equation, referred to often in biochemistry texts, can be derived.

$$pH = pK_a + \log \frac{[\text{anion, } A^-]}{[\text{acid, HA}]}$$

Similarly, the OH^- concentration in a basic buffer can be calculated by solving the equilibrium constant expression for $[OH^-]$,

$$[OH^-] = K_b \frac{[B]}{[HB^+]}$$

and

$$pOH = pK_b + \log \frac{[\text{cation, } BH^+]}{[\text{base, B}]}$$

There is really no need to memorize these equations, however, because the results can always be obtained by applying the principles learned in Section 15.2 (Example 15.8, for instance).

Review Example 15.12 (which shows you how to determine the acid-to-salt ratio needed to give a desired pH) and Example 15.13 (which shows you how to calculate the effect of additions of strong acid or base to a buffer).

Remember that when a salt like NH_4Cl or $NaC_2H_3O_2$ is dissolved in water it is completely dissociated. If the concentration of NH_4Cl in a solution is 0.10 M, the concentration of NH_4^+ (from this salt) is also 0.10 M. If the equilibrium involves NH_3 and NH_4^+, the Cl^- doesn't matter. It is there only to keep the solution electrically neutral; almost any anion would do. The Cl^- is a spectator ion.

Self-Test

13. Calculate the pH of the following solutions:

(a)   0.25 M $HC_2H_3O_2$, 0.15 M $NaC_2H_3O_2$ _____

(b)   0.25 M NH_3, 0.15 M NH_4Cl _____

14.   What ratio of formate ion (CHO_2^-) to formic acid ($HCHO_2$) must be maintained to give a solution with a pH = 3.50 (for $HCHO_2$, $K_a = 1.8 \times 10^{-4}$)?

_____

15.   How much will the pH change if 0.2 mol of HCl is added to 1.0 liter of a buffer composed of 1 M $HCHO_2$ and 1 M $NaCHO_2$ (the K_a is given in Question 14 above)?

_____

New Terms

buffer

15.5 HYDROLYSIS

Objectives

To consider the equilibria that result when salts of weak acids or bases are dissolved in water.

Review

Hydrolysis consists of the reaction of a substance with water. Several possibilities are dealt with in this section:

(1)   Salts of a weak acid and a strong base - An example is $NaC_2H_3O_2$. Only the anion hydrolyzes.

$$A^- + H_2O \rightleftharpoons HA + OH^-$$

The resulting solution is slightly basic. Remember that $K_h = K_w/K_a$.

(2)   Salts of a strong acid and a weak base - An example is NH_4Cl. Only the cation hydrolyzes.

$$HB^+ + H_2O \rightleftharpoons B + H_3O^+$$

The resulting solution is slightly acidic. Remember that $K_h = K_w/K_b$.

Numerical problems dealing with hydrolysis equilibria of these two types are illustrated in Examples 15.14 and 15.15 in the text. Compare the method of solution of these examples with that of Example 15.1 in the Study Guide. Except for the evaluation of the hydrolysis constant, they are really the same. We end up with an expression,

$$\frac{(x)(x)}{(concentration)} = (equilibrium\ constant)$$

(3) Salts of a weak acid and a weak base - An example is $NH_4C_2H_3O_2$. In this case both the cation and anion hydrolyze. If you have K_a and K_b for the weak acid and base, you can calculate K_h for each of them. If the K_h for the anion is larger than K_h for the cation, the solution is basic. An acidic solution results if K_h for the cation is greater than K_h for the anion.

(4) Salts of polyprotic acids - An example is Na_2CO_3. Remember that you only have to consider the first step in the hydrolysis. If the anion is from a diprotic acid, remember that K_h is calculated using K_{a2} for the second step in the dissociation of the acid. Review Example 15.16.

Self-Test

16. Calculate the H^+ concentration in 0.10 M NaF. For HF, $K_a = 6.5 \times 10^{-4}$.

_____

17. Calculate the H^+ concentration in 0.10 M CH_3NH_3Cl. For CH_3NH_2, $K_b = 3.7 \times 10^{-4}$.

_____

18. Will a solution of NH_4NO_2 be acidic or basic? $K_a = 4.5 \times 10^{-4}$ for HNO_2, $K_b = 1.8 \times 10^{-5}$ for NH_3.

_____

19. Calculate the H^+ concentration in a 0.050 M solution of K_2SO_3. For H_2SO_3, $K_{a1} = 1.5 \times 10^{-2}$, $K_{a2} = 1.0 \times 10^{-7}$.

_____

New Terms

hydrolysis
hydrolysis constant

15.6   ACID-BASE TITRATIONS: THE EQUIVALENCE POINT

Objectives

To see how the pH at the equivalence point in an acid-base titration is influenced by hydrolysis. You should be able to calculate the pH at various points during a titration.

Review

Three situations are discussed in this section:

(1) <u>Titrations of strong acid-strong base</u>. When base is added to the acid, complete neutralization occurs and the H^+ concentration is decreased in direct proportion to the amount of OH^- added.

(2) <u>Titration of a weak acid-strong base</u>. At the start of the titration the pH is controlled by the presence of the weak acid. Example 15.1 in the Study Guide illustrated how the H^+ concentration could be obtained for a solution of a weak acid.

  After some base has been added, the amount of weak acid has decreased and some of the anion of the acid is generated. In Section 15.4 you saw that a buffer can be prepared as a mixture of a weak acid and its anion. Therefore, between the start of the titration and the equivalence point the pH is calculated in the same way you would compute the pH of a buffer solution.

  At the equivalence point the weak acid has been completely neutralized; that is, it has been converted to a salt of the acid. We have just seen in the last section that the anion of a weak acid hydrolyzes. As a result, calculation of the pH at the equivalence point is a hydrolysis problem.

(3) <u>Titration of a weak base-strong acid</u>. The computations involved here are essentially identical to those for the weak acid-strong base.

Self-Test

20. Consider the neutralization of 0.10 M HOCl by the addition of solid NaOH (so that no volume change occurs). Calculate:

    (a) the pH before any NaOH is added   _____

    (b) the pH when half the HOCl has been
         neutralized   _____

    (c) the pH at the equivalence point   _____
    (For HOCl, $K_a = 3.1 \times 10^{-8}$ M)

New Terms

15.7 ACID-BASE INDICATORS

Objectives

To learn how an acid-base indicator functions.

Review

An acid-base indicator is itself a weak acid (or base). The molecular form of the indicator, HIn, has one color and the ionic form, In^-, has a different color. The color that is observed in a solution of the indicator is controlled by the ratio of [HIn] to [In^-], which is determined in turn by the H^+ concentration in the solution.

Self-Test

21. The indicator, methyl orange, has pK = 3.5. In acid the indicator is red; in base it is yellow-orange. What color will a solution of methyl orange be if the [H^+] = 2.1×10^{-3} M?

_____

New Terms

indicator

Answers to Self-Test Questions

1.(a) 1.0×10^{-6} (b) 2.4×10^{-3} M (c) 1.2×10^{-13} M

2.(a) 1.0×10^{-10} M (b) 3.6×10^{-6} M (c) 1.5×10^{-3} M

3.(a) pH = 5.70, pOH = 8.30 (b) pH = 8.07, pOH = 5.93

(c) pOH = 5.47, pH = 8.53 4.(a) 3.2×10^{-9} M (b) 4.6×10^{-14}

M (c) 2.2×10^{-1} M 5. 1.4×10^{-5} 6. 4.4×10^{-4} (Did you re-

member to obtain [OH$^-$] from pOH?) 7. [OH$^-$] = 1.9×10^{-6} M,

[H$^+$] = 5.3×10^{-9} M 8. 4.4×10^{-3} % 9. pK_a = 9.31

10. pH = 4.04 11. [H$^+$] = [HA$^-$] = 4.5×10^{-3} M, [A^{2-}] = K_{a2} =

5.0×10^{-9} M 12. 1.3×10^{-5} M 13.(a) 4.52 (b) 9.48

14. [CHO$_2^-$]/[HCHO$_2$] = 0.56 15. pH decreases by 0.17

16. [OH$^-$] = 1.2×10^{-6} M, [H$^+$] = 8.1×10^{-9} M 17. [H$^+$] =

1.6×10^{-6} M 18. K_h for NO$_2^-$ = 2.2×10^{-11}, K_h for NH$_4^+$ =

5.6×10^{-10}; the solution will be acidic. 19. [OH$^-$] =

7.1×10^{-5} M, [H$^+$] = 1.4×10^{-10} M 20.(a) 4.25 (b) pH =

pK_a = 7.51 (c) 10.25 21. red, since [HIn] > [In$^-$]

16 SOLUBILITY AND COMPLEX ION EQUILIBRIA

This brief chapter concludes the discussion on the quantitative aspects of ionic equilibria by considering equilibria involving salts of low solubility. We will also examine equilibria of complex ions and the way that solubility is affected by complex ion formation.

16.1 SOLUBILITY PRODUCT

Objectives

To deal quantitatively with the solubility equilibria involving "insoluble" salts. You should learn how to write the equilibrium constant expression for the solubility equilibrium. You should be able to calculate K_{sp} from solubility, and solubility from K_{sp}. You should learn how to use the K_{sp} expression to predict whether or not a precipitate will form in a given solution.

Review

Remember that salts are completely dissociated when dissolved in water. The K_{sp} expression involves only the product of ion concentrations raised to exponents that are the coefficients in the chemical equation for the equilibrium.

Problems dealing with K_{sp} can be divided into three classes:

(1) <u>Calculation of K_{sp} from solubility</u> - This is shown in Examples 16.1 and 16.2 in the text. Calculate the ion concentrations from the molar solubility and the balanced equilibrium equation. For example, the molar solubility of Ag_2CrO_4 is 7.8×10^{-5} M. The equilibrium is

$$Ag_2CrO_4 \rightleftharpoons 2Ag^+ + CrO_4^{2-}$$

From the stoichiometry of the equation we would conclude that if 7.8×10^{-5} mol/liter of Ag_2CrO_4 dissolves, then $[Ag^+] = 2(7.8 \times 10^{-5}$ M$) = 1.6 \times 10^{-4}$ M and $[CrO_4^{2-}] = 7.8 \times 10^{-5}$ M.

(2) <u>Calculation of solubility from K_{sp}</u> - This is illustrated in Examples 16.3 and 16.4. The important thing to remember with this kind of problem is to work through it methodically. Don't try to skip any of the reasoning steps. If you do, you are likely to make mistakes. Review the reasoning in these examples to be sure you understand it. Note that in constructing the concentration table for the problem the coefficient preceding the x's in the table are the <u>same</u> as the coefficients in the balanced equation for the equilibrium.

(3) <u>Determining whether precipitation will occur</u> - Determine the concentrations of the ions in the solution in question. Then remember the following:

$$\left. \begin{array}{l} \text{ion product} < K_{sp} \\ \text{ion product} = K_{sp} \end{array} \right\} \quad \text{no precipitate will form}$$

ion product $> K_{sp}$ precipitate will form

Keep in mind that if the final solution is formed by mixing two solutions, you must consider dilution. Each solute is diluted when the other solution is added. Review Examples 16.5 and 16.6 in the text.

In Example 16.7 note that we can control the concentration of the precipitating ion (S^{2-} in this example), which comes from a weak acid, by properly adjusting the pH of the solution. This example is interesting because it demonstrates how ions can be separated by selective precipitation where the concentration of the precipitating agent (S^{2-} in this case) is controlled by the pH of the solution.

Self-Test

1. The molar solubility of CuCl is 5.7×10^{-4} M. What is the value of K_{sp} for CuCl?

   _____

2. The molar solubility of PbBr$_2$ is 1.05×10^{-1} M. What is the value of K_{sp}?

   _____

3. $K_{sp} = 7.0 \times 10^{-10}$ for SrCO$_3$. What is the molar solubility of SrCO$_3$?

   _____

4. $K_{sp} = 2 \times 10^{-33}$ for Al(OH)$_3$. What is the Al^{3+} concentration in a saturated solution of Al(OH)$_3$?

   _____

5. A solution is prepared by mixing 100 ml of 0.10 M LiCl solution with 200 ml of 0.30 M NaF solution. LiF has $K_{sp} = 5 \times 10^{-3}$. Will a precipitate form in this solution?

   _____

6. A solution of 0.010 M Ca^{2+} also contains 0.10 M HF. The solution has had its acidity adjusted to a pH = 2.00 by addition of HCl. Given that for CaF$_2$, $K_{sp} = 1.7 \times 10^{-10}$ and for HF, $K_a = 6.5 \times 10^{-4}$, will a precipitate of CaF$_2$ form in this solution?

   _____

7. A solution contains 0.10 M Ca^{2+} and 0.10 M Ba^{2+}. HF is to be added until its concentration is 0.40 M to selectively precipitate CaF$_2$. What range of H$^+$ concentrations in the solution will permit CaF$_2$ to begin to precipitate without precipitating any BaF$_2$? For CaF$_2$, $K_{sp} = 1.7 \times 10^{-10}$; for BaF$_2$, $K_{sp} = 1.7 \times 10^{-6}$; for HF, $K_a = 6.5 \times 10^{-4}$.

   _____

New Terms

solubility product constant
molar solubility
ion product

16.2   COMMON ION EFFECT AND SOLUBILITY

Objectives

To see how the solubility of a substance is decreased by the presence of salts that provide a "common ion."

Review

The addition of a common ion to a solution containing a salt in equilibrium with its ions decreases the solubility of the salt. Review Examples 16.8, 16.9 and 16.10 in the text. Once again, approach this kind of problem in a deliberate, stepwise fashion; don't try to rush or skip steps in the reasoning.

Example 16.9 is typical of the kind of question that so often makes students ask, "When am I supposed to double something and when don't I double something?" Examine the concentration table constructed for this problem. Notice that in the column labeled "Initial concentration" the concentrations of the OH^- from the NaOH, which is already dissolved in the solution, is entered without reference to the equation for the equilibrium. If a solution contains 0.10 M NaOH, the OH^- concentration <u>from this source</u> is 0.10 M, not double that value. Now look at the column headed "Change." This is the column that contains x; the x's have coefficients corresponding to the coefficients of the ions in the balanced equation for the equilibrium. Remember, the only thing ever doubled (or tripled, etc.) are entries in the "Change" column.

Self-Test

8. $K_{sp} = 2 \times 10^{-8}$ for $PbSO_4$. What is the molar solubility of $PbSO_4$ in 0.010 M Na_2SO_4?

   _____

9. $K_{sp} = 2 \times 10^{-15}$ for $Fe(OH)_2$. What is the molar solubility of $Fe(OH)_2$ in a solution having a pH of 10.0?

   _____

10. $K_{sp} = 4.5 \times 10^{-17}$ for $Zn(OH)_2$. What is the molar solubility of $Zn(OH)_2$ in 1.0×10^{-3} M $Ba(OH)_2$?

   _____

11. What is the molar solubility of $Zn(OH)_2$ in 1.0×10^{-3} M

$Zn(NO_3)_2$? _____

New Terms

common ion effect

16.3   COMPLEX IONS

Objectives

To learn what complex ions are and to examine their equilibria. You should learn how to write expressions for the formation constant and the instability constant for a complex ion.

Review

The ions or molecules that attach themselves to the central atom in a complex ion are called ligands. The formation constant (or stability constant) is the K for the reaction in which the complex ion appears as a product (i.e., a formation reaction). For example,

$$Cu^{2+} + 4Cl^{-} \rightleftharpoons CuCl_4{}^{2-}$$

$$K_{form} = \frac{[CuCl_4{}^{2-}]}{[Cu^{2+}][Cl^{-}]^4}$$

The instability constant is the reciprocal of K_{form}.

Self-Test

12.   Write expressions for K_{form} and K_{inst} for the complex ion $Zn(OH)_4{}^{2-}$.

New Terms

complex ion   formation constant
ligand   stability constant
instability constant

16.4   COMPLEX IONS AND SOLUBILITY

Objectives

> To see how the formation of complex ions can affect
> solubility.

Review

The formation of complex ions affects the solubility of
some salts through a system of simultaneous equilibria, where
shifting the position of one equilibrium affects the concentration
of a species that is also involved in another equilibrium.   This
then shifts the position of the second equilibrium.   The overall
equilibrium constant for the reaction, K_c, is the product of the
K_{sp} for the insoluble salt and K_{form} of the complex ion.

$$K_c = K_{sp} \cdot K_{form}$$

Review the sample calculations in Examples 16.11 and 16.12.

Self-Test

13.   How many moles of $Cu(OH)_2$ will dissolve in 1.0 liter of
      1.0 M NH_3? $K_{sp} = 1.6 \times 10^{-19}$ for $Cu(OH)_2$; $K_{inst} =$
      2.1×10^{-13} for $Cu(NH_3)_4^{2+}$.   Ignore the ionization of NH_3
      as a weak base.

      _____

New Terms

Answers to Self-Test Questions

1. 3.2×10^{-7} 2. 4.63×10^{-3} 3. 2.6×10^{-5} 4. 2.9×10^{-9} M

5. yes; $[Li^+][F^-] = (3.3 \times 10^{-2}$ M$)(2.0 \times 10^{-1}$ M$) = 6.6 \times 10^{-3} >$

K_{sp} 6. yes; $[F^-] = 6.5 \times 10^{-3}$, ion product $= 4.2 \times 10^{-7} > K_{sp}$

7. $[H^+]$ must be less than 6.3 M and greater than or equal to

0.063 M.

8. 2×10^{-6} M

9. 2×10^{-7} M

10. 1.1×10^{-11} M

11. 1.0×10^{-7} M

12.

$$K_{form} = \frac{[Zn(OH)_4^{2-}]}{[Zn^{2+}][OH^-]^4} \qquad K_{inst} = \frac{[Zn^{2+}][OH^-]^4}{[Zn(OH)_4^{2-}]}$$

13. 5.7×10^{-3} mol

17 ELECTROCHEMISTRY

This chapter examines the way redox reactions can be caused to occur by the action of electricity, and the way electricity can be obtained from redox reactions that occur spontaneously. The study of electrochemical properties has a wide-range of applications, from the construction of new and improved batteries and fuel cells to the study of the electrolyte balance in living cells.

Several new terms are introduced in the introduction to this chapter. They are: electrolytic cell, galvanic cell, voltaic cell, electrolysis.

17.1   METALLIC AND ELECTROLYTIC CONDUCTION

Objectives

To compare the way electrical charge is transported in metals and in solutions of electrolytes.

Review

In metals, conduction occurs by the movement of electrons; in solutions of electrolytes conduction takes place by the movement of ions. Remember that in every microscopic portion of the solution electrical neutrality is maintained, as illustrated in Figure 17.2.

New Terms

metallic conduction
electrolytic conduction

17.2   ELECTROLYSIS

Objectives

To examine the processes that take place at the cathode
and anode in an electrolytic cell. You should learn from
the examples given how the net reaction in the electrolysis
of aqueous solutions is controlled by which redox reactions
occur most easily.

Review

In any electrochemical cell, we define the electrodes as:

cathode - electrode where reduction occurs
anode   - electrode where oxidation occurs

The net reaction in a cell is the sum of cathode (reduction) and
anode (oxidation) half-reactions. The sum is taken so that an
equal number of electrons are gained and lost. This is the same
procedure you learned in the ion-electron method of balancing
equations.

During the electrolysis of aqueous solutions there are usually competing reactions. The half-reactions that occur are those
that take place most easily. Review the discussions on the electrolysis of aqueous NaCl, $CuSO_4$, $CuCl_2$, and Na_2SO_4.

Self-Test

1. What is the role of Na_2SO_4 in the electrolysis of aqueous
    Na_2SO_4 solutions?

    _____

2. Write the anode and cathode reactions for the electrolysis of
    H_2O.

    _____   _____

3. In the electrolysis of brine (aqueous NaCl), Cl_2 is produced
    at the anode rather than O_2. What does this imply about the

relative ease of oxidation of H_2O and Cl^-?

_____

New Terms

electrolysis   anode

cathode   cell reaction

17.3   PRACTICAL APPLICATIONS OF ELECTROLYSIS

Objectives

To see how electrolysis is applied in industrial processes that affect the way we live.

Review

You should become familiar with the processes involved in the commercial electrolysis of sodium chloride (both molten and aqueous), the preparation of aluminum and magnesium, and the purification (refining) of copper.

Self-Test

4. What is the function of the cryolite in the Hall process?

_____

5. Write chemical equations for the separation of magnesium from sea water.

_____

_____

_____

6. What reaction occurs at the cathode in the electrolytic refining of copper?

_____

7. Write equations for the electrode reactions in the electrolysis of aqueous NaCl using the mercury cell.

anode: _____

cathode:_____

8. Why is the mercury cell used in the electrolysis of aqueous
    NaCl?

    _____

    _____

9. If an object were to be electroplated with tin from a solution
    of $SnCl_2$, would the object be made the anode or the cathode?
    Why?

    _____

    _____

New Terms

Hall process   electroplating
cryolite

17.4   QUANTITATIVE ASPECTS OF ELECTROLYSIS

Objectives

To examine, for electrolysis, the quantitative relationships
between the amount of electricity consumed and the amount
of chemical change produced. You should be able to perform calculations of the type illustrated in this section.

Review

The important relationships used in this section are:

$$1 \text{ faraday } (\mathscr{F}) = 1 \text{ mol of electrons}$$
$$1\mathscr{F} = 96,500 \text{ coulombs (C)}$$
$$1 \text{ coulomb } = 1 \text{ ampere x } 1 \text{ second}$$
or
$$1 \text{ C} = 1 \text{ A} \cdot \text{s}$$

Review Examples 17.1 and 17.2, which illustrate how these relationships are used. Remember that you must have a balanced
half-reaction or know the number of electrons transferred in order to solve these problems.

A coulometer is a device in which the amount of chemical
change produced in one cell is used to determine the number of
faradays that have passed through another cell connected to it
in series. Review Example 17.3.

Self-Test

10. How many (a) coulombs and (b) faradays are supplied by a current of 10.0 A for 8.00 hours?

(a) _____   (b) _____

11. How many grams of Cr will be produced by reduction of Cr^{3+} with a current of 1.50 A for 30.0 minutes?

_____

12. How many hours must a current of 14.0 A flow to reduce 1.00 mol of Al^{3+} to metallic aluminum in the Hall process?

_____

13. How many grams of Al are deposited on an electrode when 14.0 g of Ag are also produced when the two cells are connected in series?

_____

14. What current would be necessary to oxidize 1.00 g of water in 2.00 hours?

_____

New Terms

faraday
coulometer

17.5   GALVANIC CELLS

Objectives

To see how electricity can be provided by a spontaneous redox reaction if the oxidation and reduction half-reactions can be physically separated.

Review

Remember that the redox reactions in a galvanic cell are separated into half-cells so that the electron transfer occurs through an external circuit. The cell compartments must be connected by a salt bridge or porous partition so that electrical neutrality can be maintained.

In a galvanic cell the anode (where oxidation occurs) is negative; the cathode is positive.

New Terms

galvanic cell   salt bridge
half-cell

17.6   CELL POTENTIALS

Objectives

To define a quantity that is a measure of the driving force of the redox reaction in a galvanic cell.

Review

The force with which a galvanic cell tends to push electrons through an external circuit is called its emf (electromotive force) and is measured in volts (V). It is also called the cell potential. Standard cell potentials, $\mathscr{E}°$, are used when all species are at unit activity and the temperature is 25°C. Cell potentials should be measured with a potentiometer or other similar device which does not draw any current from the cell while the measurement is being made.

Remember that the volt is a measure of the energy that is delivered by a flowing current.

$$1 \text{ volt} = 1 \text{ joule/coulomb}$$
$$1 \text{ V} = 1 \text{ J/C}$$

Self-Test

15. What three factors influence the cell potential for a redox reaction?

_____

_____

16. What is observed if one attempts to measure a cell potential with an ordinary voltmeter instead of a potentiometer?

_____

17. How much work is done by the flow of 1.20 A for 5.00 min
     under a potential of 110 V?   _____

New Terms

electromotive force   cell potential
emf   standard cell potential
volt   potentiometer

17.7   REDUCTION POTENTIALS

Objectives

To treat the observed cell potential as the difference between the potentials of competing reduction reactions.
Also, to devise a system for tabulating standard reduction
potentials for a series of half-reactions. You should learn
how to use standard reduction potentials to calculate the
cell potential for an overall reaction. You should also
learn how to predict the spontaneity of a reaction.

Review

In this section the idea is developed that the measured
cell potential can be viewed as the difference between two reduction potentials. The reduction potential is a measure of the tendency of a reduction half-reaction to occur. When the reaction occurs at 25°C with all species at unit concentration, the term
standard reduction potential is used.

The standard hydrogen electrode is assigned a reduction
potential of exactly 0.000 volts. Other reduction potentials are
compared to that of the hydrogen electrode. A positive $\mathscr{E}°$ means
a half-reaction has a greater tendency to occur than the reaction,

$$2H^+ + 2e^- \rightleftharpoons H_2(g)$$

For a given overall reaction that can be divided into half-reactions, the standard cell potential is obtained as,

$$\mathscr{E}°_{cell} = (\mathscr{E}°_{substance\ reduced}) - (\mathscr{E}°_{substance\ oxidized})$$

Review the list of uses to which the table of reduction potentials (Table 17.1) can be put. Remember that a spontaneous reaction occurs only if the calculated $\mathscr{E}^\circ_{cell}$ is positive. Learn the diagonal relationship described under $\underline{3}$ on Page 540.

Self-Test

18. Calculate the value of $\mathscr{E}^\circ_{cell}$ for the following reactions.

(a) $Sn^{2+} + H_2 + 2\,OH^- \longrightarrow 2H_2O + Sn$ _____

(b) $4Fe + 3\,O_2 + 12H^+ \longrightarrow 6H_2O + 4Fe^{3+}$ _____

(c) $2Al + 3Zn^{2+} \longrightarrow 2Al^{3+} + 3Zn$ _____

19. Determine whether the following are spontaneous reactions.

(a) $Sn^{2+} + 2SO_4^{2-} \longrightarrow Sn + S_2O_8^{2-}$ _____

(b) $Pb + PbO_2 + 4H^+ + 2SO_4^{2-} \longrightarrow 2PbSO_4 + 2H_2O$

_____

(c) $Mn^{2+} + 2Cl^- \longrightarrow Cl_2 + Mn$ _____

20. What is the cell potential and the spontaneous reaction that occurs between the following two half-reactions?

$2NO + 2H^+ + 2e^- \rightleftharpoons H_2N_2O_2$ $\mathscr{E}^\circ = +0.71$ V

$Rh^{3+} + 3e^- \rightleftharpoons Rh$ $\mathscr{E}^\circ = +0.80$ V

_____

New Terms

reduction potential   standard hydrogen electrode
standard reduction potential

17.8   SPONTANEITY OF OXIDATION-REDUCTION REACTIONS

Objectives

To relate the standard potential for a cell to ΔG° for the cell reaction.

Review

Remember the equations,

$$\Delta G = -n\mathcal{F}\mathscr{E}$$

and
$$\Delta G° = -n\mathcal{F}\mathscr{E}°$$

where n is the number of moles of electrons transferred and $\mathcal{F}$ is the faraday constant.

Self-Test

21. The standard cell potential for the reaction,

$$Mg + Ni^{2+} \longrightarrow Ni + Mg^{2+}$$

   is 2.13 V. Calculate $\Delta G°$ for the reaction. _____

22. What is $\Delta G°$ for the reaction in Self-Test Question 20?

   _____

23. What would be the cell potential for a reaction that had reached a state of chemical equilibrium? Explain your answer? _____

New Terms

17.9   THERMODYNAMIC EQUILIBRIUM CONSTANTS

Objectives

To relate the standard cell potential to the thermodynamic equilibrium constant.

Review

Learn the equation,

$$\mathscr{E}° = \frac{0.0592}{n} \log K_c$$

This equation can be used to evaluate K_c from measured standard cell potentials. Review Example 17.5 in the text.

24. Using the data in Table 17.1, evaluate the thermodynamic equilibrium constant for the reaction,
$2Fe^{2+} + Br_2 \rightleftharpoons 2Fe^{3+} + 2Br^-$ _____

25. What is the equilibrium constant for the reaction in Self-Test Question 20?   _____

New Terms

17.10 CONCENTRATION EFFECT ON CELL POTENTIAL

Objectives

To obtain a relationship between the cell potential and the concentrations of the species involved in the cell reaction.

Review

Learn the Nernst equation,

$$\mathscr{E} = \mathscr{E}^\circ - \frac{0.0592}{n} \log Q$$

where Q represents the mass action expression for the cell reaction (omitting the concentrations of any pure solids), and n is the total number of electrons transferred. Thus, the Nernst equation for the reaction

$$4Zn(s) + NO_3^-(aq) + 10\,H^+(aq) \longrightarrow NH_4^+(aq) + 4Zn^{2+}(aq) + 3H_2O$$

in which 8 electrons are transferred is

$$\mathscr{E} = \mathscr{E}^\circ - \frac{0.0592}{8} \log \frac{[NH_4^+][Zn^{2+}]^4}{[NO_3^-][H^+]^{10}}$$

Notice that common logs are used, not natural logs.

Self-Test

26. Calculate the potential generated by the cell reaction,

    $2Al + 3Fe^{2+}(0.0010 \text{ M}) \longrightarrow 2Al^{3+}(0.10 \text{ M}) + 3Fe$ _____

27. A cell is constructed in which one electrode consists of a Zn electrode dipping into 1.0 M $ZnSO_4$. The other electrode consists of a silver electrode in a solution containing Ag^+ of unknown concentration. The cell potential is observed to be 1.40 V with the Zn serving as the anode.

    (a) What is the cell reaction? _____

    (b) What is the Ag^+ concentration? _____

New Terms

Nernst equation

17.11 APPLICATIONS OF THE NERNST EQUATION

Objectives

To learn how the Nernst equation can be applied to electrochemical measurements of different kinds.

Review

For a concentration cell, which consists of two half-cells constructed of the same substances but with different concentrations of the electrolyte,

$$\mathscr{E}_{cell} = -\frac{0.0592}{n} \log \frac{[M^{n+}]_{dil}}{[M^{n+}]_{conc}}$$

Being able to use a measured cell potential to compute the concentration of an ion has a number of applications, including the computation of K_{sp} and the monitoring of the pH of a solution.

Self-Test

28. Calculate the potential of the concentration cell consisting of one iron electrode immersed in a 0.0010 M Fe^{3+} solution and another iron electrode immersed in a 0.10 M Fe^{3+} solution.

29. Gaseous HCl was added to a solution of $AgNO_3$, causing AgCl to precipitate, until the Cl^- concentration in the solution was 0.10 M.   A silver electrode was immersed in the solution and this half-cell was then connected to a zinc half-cell containing 1.00 M Zn^{2+}.   The measured cell potential was 1.04 V with Zn serving as the anode.

   (a) What is the cell reaction? _____

   (b) What is the Ag^+ concentration in the solution containing the Ag electrode?

      _____

   (c) Compute K_{sp} for AgCl from these data.

      _____

New Terms

concentration cell

17.12   ION-SELECTIVE ELECTRODES

Objectives

   To describe some applications of galvanic cells that respond to changes in the concentration of a single ion.

Review

   This section is to provide you with an awareness of the scope of applications of ion-selective electrodes.

New Terms

ion-selective electrode
glass electrode
enzyme-substrate electrode

17.13   SOME PRACTICAL GALVANIC CELLS

Objectives

   To examine the chemistry of common (and not so common)

practical galvanic cells that are used as sources of electricity.

Review

Learn the chemical reactions that occur at the cathode and anode in the zinc-carbon dry cell, the alkaline battery, the silver oxide battery, the lead storage battery, and the nickel-cadmium cell. Review the principle of operation of the fuel cell as well as its potential advantages over conventional power sources.

Self-Test

30. After having reviewed this last section, write the chemical equations for the cathode and anode reactions in (a) the zinc-carbon dry cell, (b) the alkaline battery, (c) the silver oxide battery, (d) the lead storage battery, and (e) nickel-cadmium cell.

(a) _____

_____

(b) _____

_____

(c) _____

_____

(d) _____

_____

(e) _____

_____

New Terms

zinc-carbon dry cell
alkaline battery
silver oxide battery

lead storage battery
nickel-cadmium cell
fuel cell

Answers to Self-Test Questions

1. It maintains electrical neutrality.
2. anode: $2H_2O \longrightarrow O_2 + 4H^+ + 4e^-$
   cathode: $2H_2O + 2e^- \longrightarrow H_2 + 2OH^-$ 3. H_2O is more difficult to oxidize than Cl^-.   4. It lowers the melting point of Al_2O_3.
5. $Mg^{2+} + 2OH^- \longrightarrow Mg(OH)_2(s)$
   $Mg(OH)_2 + 2HCl \longrightarrow MgCl_2 + 2H_2O$
   $MgCl_2(\ell) \longrightarrow Mg(\ell) + Cl_2(g)$
6. $Cu^{2+}(aq) + 2e^- \longrightarrow Cu(s)$
7. anode: $Na^+(aq) + e^- \longrightarrow Na$ (in Hg)
   cathode: $2Cl^-(aq) \longrightarrow Cl_2(g) + 2e^-$
8. The NaOH produced by the net overall process is not contaminated with NaCl.   9. Cathode, because Sn would have to be reduced from Sn^{2+} to $Sn°$.   10.(a) 2.88×10^5 C   (b) $2.98 \mathscr{F}$
11. 0.485 g   12. 5.74 hr   13. 1.17 g Al   14. 1.49 A   15. nature of the species involved, their concentration, temperature
16. The voltage is lower when measured with a voltmeter.
17. 1.20 A x 5.00 min x 60 s/min = 360 C; 360 C x 110 J/C = 39.6 kJ   18.(a) 0.69 V   (b) 1.27 V   (c) 0.91 V   19.(a) not spont. (b) spont.   (c) not spont.   20. $\mathscr{E}° = 0.09$ V;
$2Rh^{3+} + 3H_2N_2O_2 \longrightarrow 6NO + 6H^+ + 2Rh$ 21. -411 kJ (-98.2 kcal)
22. -52.1 kJ (-12.5 kcal)   23. $\mathscr{E} = 0$.   At equilibrium $\Delta G = 0$. Since $\Delta G = -n\mathscr{F}\mathscr{E}$, $\mathscr{E}$ must also be zero.   24. 6.5×10^{10}
25. 1.3×10^9 26. 1.16 V
27.(a) $Zn + 2Ag^+ \longrightarrow Zn^{2+} + 2Ag$
   (b) $[Ag^+] = 2 \times 10^{-3}$ M
28. 0.039 V
29.(a) $Zn + 2Ag^+ \longrightarrow Zn^{2+} + 2Ag$
   (b) $[Ag^+] = 3 \times 10^{-9}$ M
   (c) $K_{sp} = 3 \times 10^{-10}$ M

30. See Pages 551-553 in the text.

18 CHEMICAL PROPERTIES OF THE REPRESENTATIVE METALS

As we discussed earlier, the chemical elements can be divided into three main types: metals, nonmetals, and metalloids. The metals themselves can be further divided into the representative metals (those found in the A-groups) and the transition metals (those in the B-groups and Group VIII). The properties of the A-group metals are determined by the electron population of the s and p orbitals of their outer shells; each of the inner subshells is either filled or empty. The transition metals, on the other hand, have a partially filled d subshell below the outer shell, and this gives them properties that differ in significant ways from the representative metals. For this reason, the transition metals and representative metals are discussed separately.

18.1 PREPARATION OF METALS

Objectives

To learn where metals are found in Nature and to learn what kinds of methods can be used to extract them from their compounds.

Review

Most metals are found in Nature in a combined state - that is, in compounds. Some occur in the oceans and on the ocean floor. On land they frequently occur in deposits of their carbonates, oxides, or sulfides.

Preparation of metals involves reducing the metal from a positive oxidation state to the free state. There are three methods described in this section:

(1) <u>Thermal decomposition</u>: This requires that the metal compound being decomposed have a small ΔH_f; otherwise excessively high temperatures must be reached to produce a measurable amount of the free metal. Be sure you understand the thermodynamic argument presented for this in the text. Review Examples 18.1 and 18.2 in the text.

(2) <u>Reduction using a chemical reducing agent</u>: This can only be used economically on metals of moderate activity. Two important chemical reducing agents are carbon and hydrogen.

(3) <u>Electrolytic reduction</u>: This must be used for very active metals. Halide salts are usually chosen for electrolysis because of their relatively low melting points. The production of aluminum is an exception.

Self-Test

1. What types of processes would probably be used to extract the metals from the following compounds?

   (a) PbO   _____

   (b) HgO   _____

   (c) NaCl   _____

2. Why would the reaction, $FeCl_3 + 3Na \longrightarrow Fe + 3NaCl$, be an uneconomical method for producing iron? _____

   _____

New Terms

18.2   GROUP IA: THE ALKALI METALS

Objectives

> To learn the chemical and physical properties of the
> Group IA metals and their important compounds.

Review

> You should study this section (as well as others in this
chapter) carefully in an effort to learn as much as possible. Pay
particular attention to the following:

Occurrence and preparation. Sodium and potassium are the most
important of the alkali metals. They are found in the ocean and
in deposits on land. Recovery of sodium is accomplished by electrolysis of molten NaCl in the Down's cell. The other alkali metals are normally displaced from their halides by sodium at high
temperatures.

Physical properties. Alkali metals are good conductors of heat
and electricity. They are soft and low-melting because their
metallic lattices contain only singly charged cations. The alkali
metals and their salts produce characteristic emission colors.
Learn the colors for the flame tests for Li, Na, and K.

Chemical properties and important compounds. All the alkali metals react vigorously with water (you should be able to write equations for the reactions). They have very negative reduction potentials. The reduction potential of Li is exceptionally negative
because of the very large hydration energy of Li^+.

> Alkali metals dissolve in liquid ammonia where they exist
as M^+ ions and solvated electrons. These solutions are very
powerful reducing agents.

> Alkali metals react directly with most of the nonmetals.
You should learn their reactions with O_2, and the uses for the
various kinds of oxides that are formed. Of the alkali metals,
only lithium reacts directly with nitrogen to form Li_3N.

> The most important compound of the alkali metals is NaCl,
which serves as the raw material for other sodium compounds.

The raw material for potassium compounds is KCl. Sodium compounds such as NaOH, $NaHCO_3$, and Na_2CO_3 are made from NaCl. You should learn the chemical reactions involved in the Solvay process.

<u>Self-Test</u>

3. Write a molecular equation for the reaction of

    (a) Na with O_2 _____

    (b) Li with N_2 _____

    (c) KO_2 with H_2O _____

    (d) Cs with O_2 _____

    (e) Na with F_2 _____

    (f) K with S _____

    (g) Na with H_2O _____

4. What are the common names for

    (a) NaOH? _____

    (b) $Na_2CO_3 \cdot 10H_2O$? _____

    (c) $NaHCO_3$? _____

5. What is potash? _____

6. Write the ionic equation for the principal reaction that takes place in the Solvay process.

_____

7. What color flame is produced by

    (a) potassium compounds? _____

    (b) lithium compounds? _____

    (c) sodium compounds? _____

8. Why are sodium vapor lamps so efficient? _____

_____

9. Give the formula of an alkali metal compound that is used as

    (a) a bleaching agent _____

    (b) an ingredient in glass _____

(c) a drug for treating manic depression   _____

(d) the most important commercial strong
     base   _____

(e) an oxygen source in recirculating
     breathing apparatus   _____

(f) a fertilizer   _____

New Terms

The common names of many compounds are included in this section and others. These will not be listed separately under "New Terms," but they are generally given in italics in the text. You should ask your teacher which of them you are expected to know. The important new terms given in boldface type in this section are as follows:

flame test
Solvay process

18.3   GROUP IIA: THE ALKALINE EARTH METALS

Objectives

To learn the chemical and physical properties of the Group IIA metals and their important compounds.

Review

Occurrence and preparation. The most important alkaline earth metals are calcium and magnesium. Learn the mineral sources of calcium. The chief source of magnesium is sea water. Beryl, an ore of beryllium, occurs sometimes as gem quality crystals.

Beryllium is obtained by electrolysis of $BeCl_2$ to which NaCl is added as an electrolyte. Learn how magnesium is separated from sea water to give the free metal. Beryllium and magnesium are the only alkaline earth metals that have practical uses as free metals. Learn why, and what some of these uses are.

<u>Physical properties</u>. The Group IIA metals are more dense, harder, and higher-melting than the metals of Group IA. Learn the colors of the flame tests for Ca, Sr, and Ba.

<u>Chemical properties and important compounds</u>. The alkaline earth metals are less reactive than the metals of Group IA. Only Ca, Sr, Ba, and Ra react with cold H_2O to liberate hydrogen. Magnesium is the only metal of this group that reacts with N_2.

Beryllium compounds are amphoteric and tend to be covalent. Learn the structure of $BeCl_2$. Study the reactions of Be and BeO with base.

Study the properties and uses of the oxides, hydroxides, sulfates, and carbonates of these metals. Especially important are: $CaCO_3$ and its thermal decomposition; MgO and $Mg(OH)_2$; $CaSO_4 \cdot 2H_2O$ (gypsum and plaster of paris); and barium sulfate.

<u>Self-Test</u>

10. Write equations for the reaction (if any) between

    (a) Ca and H_2O _____

    (b) Be and H_2O _____

    (c) CaO and H_2O _____

    (d) Ca and N_2 _____

11. Give formulas for the substances having the following common names:

    (a) gypsum   _____

    (b) lime   _____

    (c) epsom salts   _____

    (d) milk of magnesia _____

    (e) slaked lime   _____

    (f) limestone   _____

12. What flame color is produced by compounds of

    (a) calcium   _____

    (b) strontium   _____

    (c) barium   _____

13. What is dolomite? _____

14. Write an equation for the thermal decomposition of limestone.

_____

15. Write equations for the recovery of magnesium from sea water. (Use a separate piece of paper.)

16. (a) Which metal is used in flashbulbs? _____

(b) What safety feature is provided by tools made of beryllium-copper alloy?

_____

(c) Why doesn't calcium have many practical uses as a free metal?

_____

17. Which has a greater solubility in water

(a) $MgSO_4$ or $BaSO_4$?   _____

(b) $Mg(OH)_2$ or $Ba(OH)_2$?   _____

New Terms

In addition to many names for common chemicals, the following important terms were introduced in this section:

calcining
slaking

18.4 METALS OF GROUPS IIIA, IVA, AND VA

Objectives

To learn the properties of these metals and some of their compounds.

Review

Learn which elements in these groups are considered to be metals. Except for aluminum, they are post-transition elements and have a pseudonoble gas configuration beneath the outer shell. The heavier metals exhibit two oxidation states. Learn what they are.

Occurrence and preparation. Aluminum is a very abundant metal, but its primary ore is bauxite. Learn how bauxite is purified so that it can be reduced in the Hall process.

Tin ore is SnO_2 and is reduced to the metal with carbon, and then purified by electrolysis. Learn the allotropes of tin. The principal ore of lead is PbS which is first converted to the oxide and then reduced with carbon.

Bismuth occurs as the oxide and sulfide. Bi_2S_3 is converted to Bi_2O_3 and then reduced with carbon. Wood's metal, an alloy of bismuth, has a very low melting point and is used in the triggering mechanism of automatic sprinkler systems.

Chemical properties and compounds. Aluminum is very reactive. The metal is protected from oxidation by a tough Al_2O_3 coating. Al_2O_3 occurs in two forms, α-Al_2O_3 which is called corundum and γ-Al_2O_3. The α form is quite unreactive and several well-known gems consist of almost pure Al_2O_3.

Metallic aluminum is amphoteric (you should be able to write equations for its reaction with both acids and bases). The reaction of Al to form Al_2O_3 is very exothermic. Study the thermite reaction.

$AlCl_3$ is largely covalent. Know the structure of Al_2Cl_6 and be able to compare it to the structure of $BeCl_2$. When aluminum salts are dissolved in water the aluminum ion exists as $Al(H_2O)_6^{3+}$. Study what happens as base is gradually added to a solution of $Al(H_2O)_6^{3+}$.

Aluminum sulfate forms double salts with Na_2SO_4, K_2SO_4, and $(NH_4)_2SO_4$. These salts are called alums and have the general formula $M^+M^{3+}(SO_4)_2 \cdot 12H_2O$. Alums are formed by Cr^{3+} and Fe^{3+} as well as Al^{3+}. Sodium alum, $NaAl(SO_4)_2 \cdot 12H_2O$, is used in baking powders.

Tin forms compounds of Sn^{2+} and Sn^{4+}. Both oxidation states are relatively stable. For lead, the 2+ state is much more stable than the 4+ state. Similarly, the 3+ state of bismuth is much more stable than the 5+ state.

Tin and lead are both amphoteric. You should be able to write equations for their reaction with both acids and bases. You should also study the applications of tin and lead compounds.

Bismuth(III) compounds tend to hydrolyze in water to give

the BiO^+ ion. Sodium bismuthate, $NaBiO_3$, is an extremely powerful oxidizing agent.

Self-Test

18. What is the pseudonoble gas configuration? _____

19. Which of these ions have a pseudonoble gas configuration: Sn^{2+}, Bi^{5+}, Al^{3+}, Sn^{4+}?

    _____

20. Which is a better oxidizing agent, SnO_2 or PbO_2? _____

21. Give the three principal chemical reactions in the purification of bauxite.

    _____

    _____

    _____

22. What is alnico? _____

23. What chemical reactions are involved in
    (a) the production of Sn from SnO_2?

    _____

    (b) the production of Pb from PbS?

    _____

24. What metals are found in the following alloys?

    (a) bronze _____

    (b) solder _____

25. What is Wood's metal? _____

    _____

26. Write chemical equations for the reaction of metallic aluminum with a strong acid and with a strong base.

    _____

    _____

27. On a separate sheet of paper, sketch the structures of molecular $AlCl_3$ and $BeCl_2$.

28. Write chemical equations showing how aluminum hydroxide dissolves in both acid and base.

_____

_____

29. Give the general formula for an alum. _____

30. Which alum is used in baking powders? _____

31. What is the reaction between plumbite ion and hypochlorite ion in basic solution?

_____

32. What is the name of the BiO^+ ion? _____

33. Balance the following equation which occurs in an acidic solution: $BiO_3^- + Mn^{2+} \longrightarrow MnO_4^- + Bi^{3+}$

_____

New Terms

thermite reaction   double salt
alum

Answers to Self-Test Questions

1.(a) chemical reduction (b) thermal decomposition (c) electrolysis 2. Because Na would have to be produced by electrolysis, which is expensive.

3.(a) $2Na + O_2 \longrightarrow Na_2O_2$

(b) $6Li + N_2 \longrightarrow 2Li_3N$

(c) $2KO_2 + 2H_2O \longrightarrow 2KOH + O_2 + H_2O_2$

(d) $Cs + O_2 \longrightarrow CsO_2$

(e) $2Na + F_2 \longrightarrow 2NaF$

(f) $2K + S \longrightarrow K_2S$

(g) $2Na + 2H_2O \longrightarrow 2NaOH + H_2$

4.(a) lye or caustic soda (b) washing soda (c) baking soda
5. K_2CO_3 6. $Na^+ + Cl^- + NH_4^+ + HCO_3^- \longrightarrow NaHCO_3 + NH_4^+ + Cl^-$ 7.(a) violet (b) red (c) yellow 8. Most of the light emitted appears in the visible region of the spectrum.

9.(a) Na_2O_2 (b) Na_2CO_3 (c) Li_2CO_3 (d) NaOH   (e) KO_2

(f) $KCl \cdot MgCl_2 \cdot 6H_2O$

10.(a) $Ca + 2H_2O \longrightarrow Ca(OH)_2 + H_2$

   (b) $Be + H_2O \longrightarrow$ no reaction

   (c) $CaO + H_2O \longrightarrow Ca(OH)_2$

   (d) $Ca + N_2 \longrightarrow$ no reaction

11.(a) $CaSO_4 \cdot 2H_2O$ (b) CaO   (c) $MgSO_4 \cdot 7H_2O$ (d) $Mg(OH)_2$

(e) $Ca(OH)_2$ (f) $CaCO_3$ 12.(a) brick red   (b) crimson

(c) yellowish-green   13. $CaCO_3 \cdot MgCl_3$

14. $CaCO_3 \xrightarrow{\text{heat}} CaO + CO_2$

15. $CaCO_3 \xrightarrow{\text{heat}} CaO + CO_2$

   $CaO + H_2O + Mg^{2+} \longrightarrow Ca^{2+} + Mg(OH)_2(s)$

   $Mg(OH)_2(s) + 2HCl \longrightarrow MgCl_2(aq) + 2H_2O$

   $MgCl_2(s) \xrightarrow{\text{electrolysis}} Mg + Cl_2$

16.(a) magnesium   (b) nonsparking   (c) It reacts with moisture and oxygen.   (Its oxide coating does not protect the metal from further reaction.)   17. (a) $MgSO_4$ (b) $Ba(OH)_2$

18. $ns^2np^6nd^{10}$ (e.g., $3s^23p^63d^{10}$)   19. Bi^{5+} and Sn^{4+}

20. PbO_2

21. $Al_2O_3 + 2OH^- \longrightarrow 2AlO_2^- + H_2O$

   $AlO_2^- + H_3O^+ \longrightarrow Al(OH)_3$

   $2Al(OH)_3 \xrightarrow{\text{heat}} Al_2O_3 + 3H_2O$

22. a magnetic aluminum alloy containing iron (50%), nickel (20%), and cobalt (10%).

23.(a) $SnO_2 + C \longrightarrow Sn + CO_2$

   (b) $2PbS + 3O_2 \longrightarrow 2PbO + 2SO_2$

   $2PbO + C \longrightarrow Pb + CO_2$

24.(a) copper and tin   (b) tin and lead

25. a low-melting alloy of bismuth that also contains lead, tin, and cadmium.

26. $2Al + 6H^+ \longrightarrow 2Al^{3+} + 3H_2$

   $2Al + 2OH^- + 2H_2O \longrightarrow 2AlO_2^- + 3H_2$

27. See Pages 581 and 575 of the text.

28. $Al(OH)_3(H_2O)_3 + H^+ \longrightarrow Al(OH)_2(H_2O)_4^+ + H_2O$

    $Al(OH)_3(H_2O)_3 + OH^- \longrightarrow Al(OH)_4(H_2O)_2^- + H_2O$

29. $M^+M^{3+}(SO_4)_2 \cdot 12H_2O$

30. $NaAl(SO_4)_2 \cdot 12H_2O$

31. $Pb(OH)_4^{2-} + OCl^- \longrightarrow PbO_2 + H_2O + 2OH^- + Cl^-$

32. bismuthyl ion

33. $14H^+ + 5BiO_3^- + 2Mn^{2+} \longrightarrow 2MnO_4^- + 5Bi^{3+} + 7H_2O$

19 THE CHEMISTRY OF SELECTED NONMETALS, PART I: HYDROGEN, CARBON, OXYGEN, AND NITROGEN

Chapters 19 and 20 deal with the chemistry of the most important nonmetals. We begin here with four elements essential to living things, although this is certainly not the only place in Nature where they are found.

19.1 HYDROGEN

Objectives

To learn the chemical and physical properties of hydrogen and its compounds. You should learn the properties of hydrogen and its isotopes, its method of preparation and its uses, and how hydrogen compounds can be made.

Review

Hydrogen is the most abundant element in the universe, but little of it remains on Earth. Most of the hydrogen that is present on Earth is bound in water molecules. There are three isotopes of hydrogen: ^1_1H, ^2_1H (deuterium), and ^3_1H (tritium).

Most naturally occurring hydrogen is $_1^1H$.

Commercially, H_2 is generally obtained from natural gas, CH_4, by reaction with steam. It can also be made from coal by the water gas reaction. Electrolysis of brine is another commercial source of H_2. In the laboratory H_2 can be made by reaction of a nonoxidizing acid with a metal that has a negative reduction potential. You should study the commercial uses of hydrogen described on Pages 590 and 591.

Compounds of hydrogen. Binary compounds of hydrogen are called hydrides. Those formed with metals are ionic and contain the H^- ion. Reaction of H^- with water releases H_2. In covalent compounds, hydrogen only forms one bond. The simple covalent hydrides (Table 19.1) have formulas and structures that are easy to predict. More complex hydrides are formed by nonmetals that experience catenation - the linking of like-atoms together in chains.

Preparation of nonmetal hydrides. Two methods of preparation are discussed. One is the direct combination of hydrogen with the nonmetals. This is only successful if ΔG_f° for the compound is negative (Table 19.2). The second method of preparation is the addition of protons (H^+) to the conjugate base of the hydride.

$$X^{n-} + nH^+ \longrightarrow H_nX$$

In general, the weaker X^{n-} is as a base, the stronger must be the acid supplying the H^+. You should learn how the base strength of X^{n-} varies within the periodic table (across rows and down columns).

Hydrogen economy. The benefits of H_2 as a fuel are many, but the most significant problem is the difficulty of obtaining H_2 from its most abundant source, H_2O. Making H_2 from H_2O requires an input of energy. Among the possible sources of this energy are nuclear energy and solar energy.

Self-Test

1. (a) Give the symbols for the three isotopes of hydrogen.

———————————————————

(b) Which one is radioactive?   ———————————————

2. What is heavy water? _____

3. Give chemical equations for

    (a) the commercial preparation of H_2 from CH_4.

    _____

    (b) the laboratory preparation of H_2.

    _____

4. Why can H_2 be collected by the displacement of water?

    _____

5. What is synthesis gas? _____

6. What occurs during the hydrogenation of an unsaturated oil?

    _____

7. Complete the following equations. If there is no reaction,
    write N.R.

    (a) $Li + H_2 \longrightarrow$

    (b) $CaH_2 + H_2O \longrightarrow$

    (c) $C + H_2O \xrightarrow{1000°C}$

    (d) $AlP + H_2O \longrightarrow$

    (e) $H_2 + Cl_2 \longrightarrow$

    (f) $P + H_2 \longrightarrow$

    (g) $NaCl + H_2SO_4(conc.) \longrightarrow$

8. What is the major obstacle to the widespread use of H_2 as a
    fuel? _____

    _____

9. Why can't H_2Se be made by the direct combination of the
    elements? _____

    _____

New Terms

protium   hydrides
deuterium   hydride ion
tritium   hydrogenation
heavy water   catenation
synthesis gas   hydrogen economy

19.2 CARBON

Objectives

To learn the chemical and physical properties of carbon and its compounds. You should know the allotropic forms of carbon and the important ionic and covalent compounds of carbon.

Review

Carbon is a very common element. Besides being found in all living things, large deposits of carbon in the form of coal occur in many places. Coke is nearly pure carbon and is made by heating coal to high temperatures.

You should know the structures of diamond and graphite and how their physical properties are influenced by their structures. You should also know the properties and uses of activated carbon and carbon black.

Carbon forms two oxides, CO and CO_2. You should know their structures. Carbon monoxide is nearly nonpolar and has a low solubility in water. It can be made by burning carbon or a hydrocarbon in a limited supply of O_2, by the dehydration (removal of water from) of formic acid, and by the reaction of steam with C or CH_4. Carbon monoxide is an industrial fuel and an industrial reducing agent. It reacts with some metals to form metal carbonyl compounds in which the metal has an oxidation number of zero.

Carbon dioxide is nonpolar and moderately soluble in water. It can be made by combustion of a carbon-containing substance by the decomposition of limestone, and by the reaction of a carbonate or bicarbonate with an acid. Solid CO_2 is Dry ice, which sublimes at $-78°C$ at 1 atm. Learn the uses of CO_2 given on Page 598. Green plants consume CO_2 and produce carbohydrates during photosynthesis.

Aqueous solutions of CO_2 contain H_2CO_3, a weak diprotic acid. Carbonic acid forms two types of salts, carbonates and bicarbonates. Ground water containing dissolved CO_2 gradually dissolves limestone deposits and produces caverns. Learn how stalactites and stalagmites are formed. You should also learn what "hard water" is and how it can be "softened."

Other carbon compounds include the carbides - covalent carbides like carborundum, ionic carbides like Al_4C_3 and CaC_2, and interstitial carbides like tungsten carbide.

Hydrogen cyanide is made from ammonia and forms cyanide salts. The CN^- forms many stable complex ions. Carbon disulfide is made by combining sulfur and carbon. It is extremely flammable.

Self-Test

10. How do the structures of diamond and graphite differ?

_____

_____

11. What special property does activated carbon have? _____

_____

12. How can CO be made in small quantities in the laboratory? (Give a chemical equation.)

_____

13. What is the overall reaction for the reduction of Fe_2O_3 by CO? _____

14. What is the oxidation number of Ni in $Ni(CO)_4$? _____

15. Give the equation for the thermal decomposition of limestone.

_____

16. What is Dry ice? _____

17. Write an equation for the dissolving of limestone by water containing dissolved CO_2.

18. (a) What ions are present in hard water? _____

(b) Write an equation showing how washing soda is able to remove hardness ions.

_____

19. (a) What is the Lewis formula of the anion in CaC_2?

_____

(b) What is the reaction of CaC_2 with water?

_____

20. What is the formula for

(a) carborundum   _____

(b) hydrogen cyanide   _____

(c) tungsten carbide   _____

New Terms

metal carbonyl compound   hardness ions
Dry ice   carborundum
hard water

19.3 OXYGEN

Objectives

To learn the properties of oxygen and its principal compounds.

Review

Note the abundance of oxygen and its presence everywhere. Also note that O_2 molecules are paramagnetic, with two unpaired electrons. Oxygen is normally recovered from liquefied air. In the laboratory it can be prepared by the MnO_2-catalyzed thermal decomposition of $KClO_3$. In nature, O_2 is a product of photosynthesis.

Oxygen exists in two allotropic forms, normal O_2 and ozone, O_3. You should be able to draw the Lewis structures for O_3 and explain its role in protecting the Earth from harmful UV radiation from the sun. You should also know how nitrogen oxides and chlorofluorocarbons can damage the Earth's ozone shield.

Ionic oxides such as CaO are basic. In water, O^{2-} reacts to form OH^-. Insoluble oxides react with acids (as in the pickling of iron). Some metal oxides are amphoteric. Learn the reaction of Al_2O_3 with base.

Covalent oxides are normally formed by nonmetals and tend to be acidic. When they react with water they form oxoacids, which produce oxoanions upon neutralization. Oxygen usually forms one or two bonds; exceptions are CO and NO. Covalent oxides can be prepared by (a) direct union of the elements, (b) oxidation of a lower oxide, (c) combustion of a metal hydride, and (d) reduction of an oxoanion in a redox reaction.

Besides normal ionic oxides containing the O^{2-} ion, oxygen also forms peroxides and superoxides. Hydrolysis of peroxide ion gives H_2O_2 (you should know its structure). H_2O_2 decomposes easily to O_2 and H_2O, and is a strong oxidizing agent.

Self-Test

21. (a) What is the usual commercial source of oxygen?

   _____

   (b) Write an equation for the laboratory preparation of oxygen.

   _____

22. Draw the Lewis structures for ozone.

23. How is ozone formed in the upper atmosphere?

   _____

   _____

24. Give chemical equations showing how traces of NO can remove relatively large amounts of O_3 from the stratosphere.

   _____

   _____

25. What is pickling? _____

   _____

26. Complete and balance the following equations:

   (a) $Al_2O_3 + H^+ \longrightarrow$

   (b) $Li_2O + H_2O \longrightarrow$

   (c) $Na_2O_2 + H_2O \longrightarrow$

27. Write equations for two reactions that could be used to prepare SO_2.

   (a) _____

   (b) _____

28. Describe the structure of hydrogen peroxide.

New Terms

ozone
pickling

19.4 NITROGEN

Objectives

To learn the chemical and physical properties of nitrogen and its important compounds.

Review

Most nitrogen on Earth exists as N_2 because of the great stability of this molecule, which results from the high $N \equiv N$ bond energy. Commercially, N_2 is obtained from liquefied air. In the laboratory, N_2 can be made by warming a solution containing NH_4^+ and NO_2^-. Most nitrogen is used to make NH_3. Other uses are based on the low reactivity of N_2 and the cold temperature of liquid N_2.

Examples of ionic nitrides are Li_3N and Mg_3N_2. Learn what happens when they react with water. In the text, the covalent compounds of nitrogen are discussed in terms of the oxidation number of the nitrogen in the compounds. Pay particular attention to the following:

3- oxidation state. The compound ammonia is discussed. Learn about the Haber process - the chemical reaction and the reaction conditions. By this time, you should know how NH_3 behaves as a base in water. The amide ion, NH_2^-, is the strong conjugate base of NH_3. The NH_4^+ ion is the weak conjugate acid of NH_3. Learn how NH_3 can be made in the laboratory. You should also learn the chemistry of the Ostwald process.

2- oxidation state. Hydrazine is discussed. You should know its structure, how it is prepared, and how it behaves in water. Hydrazine is a powerful reducing agent.

1- oxidation state. Know the structure of hydroxylamine and how this compound is prepared. Be able to write an equation for the functioning of hydroxylamine as a weak base.

1+ oxidation state. You should know how N_2O is prepared and be able to draw its Lewis structure. This and other nitrogen oxides have positive standard free energies of formation. They are stable at room temperature only because they decompose at an extremely slow rate. Learn the uses of N_2O.

2+ oxidation state and 4+ oxidation state. Oxidation of NH_3 in the Ostwald process gives NO, which is easily oxidized to NO_2. In water, NO_2 reacts to give HNO_3 and NO.

Nitric oxide is a colorless, fairly reactive gas. Its bond order is 2.5 and rather easily forms the NO^+ ion, which has a bond order of 3.0. Nitrogen dioxide can be represented by two resonance structures, each of which places the unpaired electron on the nitrogen. Learn the equilibrium between NO_2 and N_2O_4, and the structure of both molecules.

You should learn the role of NO and NO_2 in photochemical smog, including (a) the formation of NO in auto engines, (b) oxidation of NO to NO_2, (c) photodecomposition of NO_2 to NO and O, (d) the formation of ozone, and (e) the oxidation of hydrocarbons to peroxyacylnitrates (PAN).

3+ oxidation state. Condensation of NO and NO_2 gives N_2O_3, the formal anhydride of HNO_2. Nitrous acid is a weak acid that can function as either an oxidizing agent or reducing agent. Nitrites are used as food preservatives. You should understand the reasons why NO_2^- may be potentially harmful as well as why it is beneficial.

<u>5+ oxidation state</u>. Molecules of N_2O_5 exist in the vapor, but in the solid N_2O_5 produces the ions $NO_2^+NO_3^-$. N_2O_5 reacts with water to form HNO_3. Nitric acid is a strong acid and a powerful oxidizing agent. It is made by the Ostwald process. In concentrated form it is subject to photochemical decomposition, which is why concentrated HNO_3 is often slightly yellow. Aqua regia is one part HNO_3 and three parts HCl by volume. Learn why it is able to dissolve noble metals.

<u>Self-Test</u>

29. Write a chemical equation for the preparation of N_2 in the laboratory.

    _____

30. Complete and balance the following equations. If no reaction occurs, write N.R.

    (a) $Li_3N + H_2O \longrightarrow$

    (b) $NO_2 + H_2O \longrightarrow$

    (c) $NO + H_2O \longrightarrow$

    (d) $NH_3 + H_2O \longrightarrow$

    (e) $CaO + NH_4Cl \longrightarrow$

    (f) $N_2O_5 + H_2O \longrightarrow$

    (g) $O_3 + NO \longrightarrow$

    (h) $Cu + HNO_3$ (dilute) $\longrightarrow$

    (i) $HNO_2 + MnO_4^-$ (in acid solution) $\longrightarrow$

    (j) $Zn + HNO_3 \longrightarrow$

    (k) $NO + NO_2 + H_2O \longrightarrow$

31. What relationship exists between NO_2 and N_2O_4?

    _____

32. (a) Write the chemical equation that occurs in the Haber process.

    _____

    (b) What is the first chemical reaction that takes place in the Ostwald process?

    _____

33. Which is the only metal that reacts with N_2 at room temperature? _____

34. Why is N_2 so unreactive? _____

_____

35. Give the Lewis structure for

   (a) N_2H_4 (d) N_2O_3

   (b) NO_2 (e) N_2O_5 (in the gas phase)

   (c) N_2O_4

36. Why do O_2 and N_2 form stable mixtures? (That is, why don't O_2 and N_2 react with each other to a large degree in the atmosphere?) _____

_____

37. Write equations for

   (a) the formation of hydrazine.

   _____

   (b) the synthesis of hydroxylamine.

   _____

   (c) the laboratory preparation of pure HNO_3.

   _____

   (d) the synthesis of nitrous oxide.

   _____

38. Give equations for the sequence of chemical reactions that lead to the formation of ozone in photochemical smog.

_____

_____

_____

_____

New Terms

Haber process   disproportionation
Ostwald process   photochemical smog

Answers to Self-Test Questions

1.(a) ^1_1H, ^2_1H, ^3_1H (b) tritium   2. deuterium oxide, $^2_1H_2O$

3.(a) $CH_4(g) + H_2O(g) \xrightarrow[\text{catalyst}]{\text{heat}} CO(g) + 3H_2(g)$

   (b) $Zn + H_2SO_4 \longrightarrow ZnSO_4 + H_2$

4. Because it has a small solubility in water.   5. A mixture of CO and H_2.   6. Hydrogen is added to carbon-carbon double bonds, converting them to single bonds.

7.(a) $2Li + H_2 \longrightarrow 2LiH$

   (b) $CaH_2 + 2H_2O \longrightarrow Ca(OH)_2 + H_2$

   (c) $C + H_2O \xrightarrow{1000°C} CO + H_2$

   (d) $AlP + 3H_2O \longrightarrow Al(OH)_3 + PH_3$

   (e) $H_2 + Cl_2 \longrightarrow 2HCl$

   (f) $P + H_2 \longrightarrow N.R.$

   (g) $NaCl + H_2SO_4 \longrightarrow HCl + NaHSO_4$

8. An economical way must be found to produce H_2 from H_2O.
9. Because H_2Se has a positive ΔG°_f.   10. Diamond has a three-dimensional network of C—C bonds; graphite consists of planar sheets of hexagonal rings.   These sheets are stacked one on the other.   11. It has a very large ratio of surface area to mass and adsorbs large quantities of molecules on its surface.

12. $HCO_2H(\ell) \xrightarrow[\text{(conc.)}]{H_2SO_4} H_2O(\ell) + CO(g)$

13. $Fe_2O_3 + 3CO \longrightarrow 2Fe + 3CO_2$ 14. zero

15. $CaCO_3 \xrightarrow{\text{heat}} CaO + CO_2$ 16. Solid CO_2

17. $CaCO_3(s) + H_2CO_3(aq) \longrightarrow Ca(HCO_3)_2(aq)$

18.(a) Ca^{2+}, Mg^{2+}, Fe^{3+}

(b) Washing soda is $Na_2CO_3 \cdot 10H_2O$, which gives CO_3^{2-} in solution. $Ca^{2+} + CO_3^{2-} \longrightarrow CaCO_3(s)$

19. (a) $[:C \equiv C:]^{2-}$ (b) $CaC_2 + 2H_2O \longrightarrow Ca(OH)_2 + C_2H_2$

20. (a) SiC (b) HCN (c) WC

21. (a) the atmosphere (extraction from liquefied air)

(b) $2KClO_3(s) \xrightarrow[\text{heat}]{MnO_2} 2KCl(s) + 3O_2(g)$

22. See Page 602.

23. $O_2 \xrightarrow{h\nu} 2O$; $O + O_2 \longrightarrow O_3$

24. $NO + O_3 \longrightarrow NO_2 + O_2$; $NO_2 + O \longrightarrow NO + O_2$

25. Removal of an oxide coating from a metal by reaction of the oxide with an acid.

26. (a) $Al_2O_3 + 6H^+ \longrightarrow 2Al^{3+} + 3H_2O$

(b) $Li_2O + H_2O \longrightarrow 2LiOH$

(c) $Na_2O_2 + 2H_2O \longrightarrow 2NaOH + H_2O_2$

27. (a) $S + O_2 \longrightarrow SO_2$

(b) $2H_2S + 3O_2 \longrightarrow 2H_2O + 2SO_2$

28. See Page 607.

29. $NH_4^+(aq) + NO_2^-(aq) \xrightarrow{\text{heat}} N_2(g) + 2H_2O$

30. (a) $Li_3N + 3H_2O \longrightarrow 3LiOH + NH_3$

(b) $3NO_2 + H_2O \longrightarrow 2H^+ + 2NO_3^- + NO$

(c) $NO + H_2O \longrightarrow$ no reaction

(d) $NH_3 + H_2O \rightleftharpoons NH_4^+ + OH^-$

(e) $CaO + 2NH_4Cl \longrightarrow CaCl_2 + 2NH_3$

(f) $N_2O_5 + H_2O \longrightarrow 2HNO_3$

(g) $O_3 + NO \longrightarrow NO_2 + O_2$

(h) $3Cu + 8HNO_3 \longrightarrow 3Cu(NO_3)_2 + 2NO + 4H_2O$

(i) $H^+ + 5HNO_2 + 2MnO_4^- \longrightarrow 5NO_3^- + 2Mn^{2+} + 3H_2O$

(j) $4Zn + 10HNO_3 \longrightarrow 4Zn(NO_3)_2 + NH_4NO_3 + 3H_2O$

(k) $NO + NO_2 + H_2O \longrightarrow 2HNO_2$

31. $2NO_2 \rightleftharpoons N_2O_4$

32.(a) $3H_2 + N_2 \rightleftharpoons 2NH_3$

   (b) $4NH_3 + 5O_2 \longrightarrow 4NO + 6H_2O$

33. Lithium

34. Because the triple bond in N_2 is so strong.

35.(a) see Page 611   (b) see Page 613   (c) see Page 614
   (d) see Page 615   (e) see Page 616

36. The ΔG_f° of the nitrogen oxides are positive, so virtually no reaction occurs.

37.(a) $2NH_3 + NaOCl \longrightarrow N_2H_4 + NaCl + H_2O$

   (b) $NaNO_2 + NaHSO_3 + SO_2 + 2H_2O \longrightarrow 2NaHSO_4 + NH_2OH$

   (c) $NaNO_2(s) + H_2SO_4(conc.) \xrightarrow{heat} NaHSO_4(s) + HNO_3(g)$

   (d) $NH_4NO_3 \xrightarrow{heat} N_2O + 2H_2O$

38. $N_2 + O_2 \longrightarrow 2NO$

   $2NO + O_2 \longrightarrow 2NO_2$

   $NO_2 \xrightarrow{h\nu} NO + O$

   $O + O_2 \longrightarrow O_3$

20 THE CHEMISTRY OF SELECTED NONMETALS, PART II: PHOSPHORUS, SULFUR, THE HALOGENS, THE NOBLE GASES, AND SILICON

In this chapter we conclude our discussion of the representative elements with an examination of the properties of a number of additional important nonmetals. As mentioned in the text, their properties cover a broad range, and as we will see, so do their uses.

20.1 PHOSPHORUS

Objectives

To learn the properties of phosphorus and its compounds, especially those containing oxygen.

Review

Phosphorus can exist in several allotropic forms. The most important are white phosphorus and red phosphorus. White phosphorus is very reactive because of the strained 60° bond angle in the P_4 tetrahedron. It is also very toxic. Red phosphorus is less reactive and relatively nontoxic. Its structure is unknown, but is probably polymeric.

Phosphorus is prepared from phosphate rock, which contains $Ca_3(PO_4)_2$, by reduction with carbon in an electric furnace.

Phosphides containing the P^{3-} ion hydrolyze to give PH_3. The most important phosphorus compounds are those with oxygen and the halogens. Two oxides are formed, P_4O_{10} and P_4O_6. You should know their structures, and how they relate to the structure of white phosphorus. P_4O_{10} is formed by burning phosphorus in an abundant supply of oxygen. It reacts with water to give H_3PO_4 and is a useful desiccant. In a limited supply of oxygen, combustion of phosphorus yields P_4O_6. This oxide reacts with water to give H_3PO_3. Phosphorus acid is a diprotic acid; be sure to learn its structure.

Phosphoric acid is made from phosphate rock by reaction with H_2SO_4, and by reaction of P_4O_{10} with H_2O. It is a triprotic acid and forms three types of salts. You should know how superphosphate fertilizer is made, how phosphate anions serve as buffers, and the uses of Na_3PO_4.

Dehydration of H_3PO_4 gives polymeric phosphates. You should know the structures of the pyrophosphate, metaphosphate, and tripolyphosphate ions, as well as their corresponding acids. Learn their uses, too. You should also know the effect that phosphate pollution has on lakes.

Phosphorus forms two series of halogen compounds, PX_3 and PX_5. You should be able to describe their structures. The most important of them are PCl_3 and PCl_5. Learn how they are made and how they react with water. PCl_5 exists as $PCl_4^+PCl_6^-$ in the solid state. It reacts with P_4O_{10} to give phosphoryl chloride, $POCl_3$.

Self-Test

1. What is the chief phosphorus-containing compound in phosphate rock?

2. Write a chemical equation showing how phosphorus is recovered from phosphate rock.

    _____

3. On a separate sheet of paper, sketch the structures of P_4, P_4O_6, and P_4O_{10}.

4. Write an equation for the reaction of sodium phosphide with water.

    _____

5. (a) Draw the Lewis structure for phosphine.

    (b) What would you predict the structure of phosphine to be?

         _____

6. Write equations for the reaction of water with

    (a) P_4O_6 _____

    (b) P_4O_{10} _____

    (c) PCl_3 _____

7. What is a desiccant? _____

    _____

8. (a) Write the formulas of all the salts that can be formed by allowing KOH to react with H_3PO_3.

         _____

    (b) How is phosphorous acid usually made?

         _____

9. Draw the Lewis structure for the pyrophosphate ion. Be sure to indicate its charge.

10. (a) Which "phosphate" is used in liquid detergents?

         _____

(b) Which one is used in solid detergents? _____

11. Sketch the structures for PCl_3 and PCl_5 on a separate piece
     of paper.

12. What is the formula for phosphoryl chloride? _____
     What is its principal use?
     _____

New Terms

electric furnace
desiccant

20.2 SULFUR

Objectives

To learn the properties of sulfur and its compounds. In
particular, you should learn the properties of elemental
sulfur and of the oxides, oxoacids, and oxoanions of
sulfur.

Review

About half of the sulfur used by industry each year is
mined by the Frasch process; the rest is recovered from natural
gas and petroleum. Several allotropes of sulfur exist. The most
stable is rhombic sulfur which contains S_8 rings. You should
know the changes that occur as sulfur is heated gradually to its
boiling point.

There are two important oxides of sulfur, SO_2 and SO_3.
Burning sulfur or compounds that contain sulfur give SO_2.
Sulfur dioxide dissolves in water to give H_2SO_3, a weak diprotic
acid that is responsible for many of the harmful effects of acid
rain. Sulfur dioxide is removed from industrial gases by reaction
with moist limestone.

Oxidation of SO_2 to SO_3 is accomplished catalytically.
When dissolved in water, SO_3 gives H_2SO_4, which is the world's
most important industrial chemical. Usually the SO_3 is dissolved
in H_2SO_4 to give $H_2S_2O_7$, which is then diluted with water to
give H_2SO_4. Concentrated H_2SO_4 is a strong dehydrating agent

and, when hot, is a reasonably strong oxidizing agent.

Sulfur forms binary sulfides with metals, many of which react with acids to give H_2S. In other "thio" compounds, sulfur replaces oxygen. Examples are thioacetamide and thiosulfate ion. Hydrolysis of thioacetamide gives H_2S. Thiosulfate ion is used in photography to dissolve undeveloped silver bromide. Oxidation of thiosulfate gives SO_4^{2-} with Cl_2 and $S_4O_6^{2-}$ with I_2.

Self-Test

13. Describe what happens when liquid sulfur is gradually heated to its boiling point.

    _____

    _____

    _____

    _____

14. How can SO_2 be conveniently prepared in the laboratory? (Write an equation.)

    _____

15. (a) What is the formula for pyrosulfuric acid? _____

    (b) What is its reaction with water?

        _____

16. Give Lewis structures for H_2SO_4 and H_2SO_3.

17. Write equations for the reaction of concentrated H_2SO_4 with iodide ion.

    _____

18. Write equations for the following reactions.

    (a) ZnS with HCl   _____

    (b) SO_2 with H_2O _____

    (c) $S_2O_3^{2-}$ with AgBr_____

(d) $S_2O_3^{2-}$ with I_2 _____

(e) S with SO_3^{2-} _____

New Terms

Frasch process

acid rain

contact process

thio

20.3   THE HALOGENS

Objectives

To learn the properties of the halogens. You should learn their relative oxidizing strengths, how the elements are prepared, and the properties of the compounds described here.

Review

The halogens always occur in the combined state in nature, primarily as the halides. Learn the sources of the individual halogens.

As free elements the halogens are diatomic (e.g., F_2, Cl_2, Br_2, I_2). Learn the trends in their boiling points and in their electronegativities. The halogens decrease in oxidizing strength from F_2 to I_2, and a particular halogen will displace one below it (in the periodic table) from its compounds. For example, Cl_2 will displace Br^- and I^- from their compounds.

Fluorine is produced by electrolysis of KF, HF mixtures. Chlorine is produced by electrolysis of molten NaCl and by electrolysis of brine. In the laboratory, Cl_2 can be made by reacting HCl with a strong oxidizing agent such as MnO_2, $KMnO_4$, or $K_2Cr_2O_7$. Bromine is recovered from sea water by reacting Cl_2 with the Br^- that is present. Iodine is obtained from seaweed and from Chilean saltpeter, a source of $NaIO_3$.

Hydrogen halides can be made by direct combination of the elements. The vigor of the reaction decreases from fluorine to iodine. In the laboratory, HF is made from CaF_2 and concentrated H_2SO_4; HCl is prepared from NaCl and concentrated H_2SO_4. Hydrogen bromide and hydrogen iodide can be made from NaBr and NaI using concentrated H_3PO_4. (You should learn why H_2SO_4 cannot be used.) HF has a relatively high boiling point because of hydrogen bonding. Water solutions of the hydrogen halides are acidic. All except HF are strong acids. HF reacts with silica to produce SiF_4.

The halogens form four types of oxoacids: HOX, HXO_2, HXO_3, and HXO_4, which each produce a corresponding oxoanion when neutralized. These are summarized in Table 20.2. You should know their names and structures. The most important are the oxoacids and oxoanions of chlorine.

Chlorine disproportionates in water to give HOCl and HCl. In cold base, OCl^- and Cl^- are formed. In hot aqueous base ClO_3^- and Cl^- are formed. Perchlorate ion is formed by disproportionation of chlorate ion. The oxoacids and oxoanions of chlorine tend to be strong oxidizing agents. Exceptions are dilute solutions of $HClO_4$, which have poor oxidizing power.

The halogens form many compounds with the other non-metals. Their formulas are determined by (a) the number of halogen atoms that must be attached to give the nonmetal a complete octet and (b) the size of the halogen atoms that are attached. When the halogen atom is large (e.g., Br or I) the maximum number that can pack around the nonmetal is influenced by crowding, and the larger the central atom, the greater is the number of these large halogens that can be accommodated.

The tendency of the nonmetal halides to hydrolyze is influenced by their electronic and molecular structures. Carbon tetrachloride has little tendency to hydrolyze because carbon has no vacant orbitals to which an attacking water molecule can become attached. In SF_6, the sulfur is protected from attack by the surrounding "cage" of fluorine atoms.

Self-Test

19. (a) What are the principal sources of fluorine?

_____

(b) What is the principal source of chlorine?_____

20. In which substance are the London forces stronger, Br_2 or Cl_2?

_____

21. Complete and balance the following equations. If no reaction occurs, write N.R.

    (a) $CaCl_2 + Br_2 \longrightarrow$

    (b) $FeBr_3 + F_2 \longrightarrow$

    (c) $PbI_2 + Cl_2 \longrightarrow$

22. Write an equation for the laboratory preparation of chlorine using MnO_2 as the oxidizing agent.

    _____

23. What is the ionic equation for the electrolysis of brine?

    _____

24. How can Br_2 be prepared in the laboratory?

    _____

25. (a) At room temperature, which halogen is a deep red liquid?

        _____

    (b) Which is a dark metallic-looking solid?_____

26. Which metal halide is likely to be most ionic, $TiCl_2$ or $TiCl_4$?

    _____

27. Write an equation for the laboratory preparation of

    (a) HF   _____

    (b) HCl   _____

    (c) HBr   _____

28. Which hydrogen halide is found in muriatic acid?_____

29. Write an equation for

    (a) the reaction of chlorine with calcium oxide.

        _____

    (b) the reaction of chlorine with cold aqueous sodium hydroxide.

        _____

    (c) the disproportionation of potassium chlorate when heated.

        _____

(d)   the disproportionation of bromine in hot aqueous base.

_____

(e)   the hydrolysis of silicon tetrachloride.

_____

30.   What is the formula of periodic acid?   _____

31.   On a separate sheet of paper, draw the Lewis structures for chlorous acid and perchloric acid.

32.   What molecular structure would be predicted for

(a) BrF_3?   _____

(b) TeI_4?   _____

(c) $GeCl_4$?   _____

33.   What reaction would you predict if $GeCl_4$ were dissolved in water?

_____

34.   What explanation is given for the ease of hydrolysis of $SiCl_4$?

_____

_____

New Terms

20.4   NOBLE GAS COMPOUNDS

Objectives

To study which of the noble gases are able to form compounds and to examine the bonding in them.

Review

The noble gases are very unreactive, but the heavier ones Kr, Xe, and Rn - have ionization energies that are sufficiently low to enable them to share electrons with other very electronegative elements. Compounds of xenon have been studied more

extensively than those of either krypton or radon (which is radioactive, and therefore difficult to work with). One of the key
points made in this section is that scientists must avoid blind
spots in their chemical thinking. Study the way the bonding in
XeF_2 and XeF_4 is explained.

Self-Test

35. What are the structures of

    (a) XeO_2? _____

    (b) XeO_4? _____

    (c) XeF_4? _____

New Terms

clathrate

20.5 SILICON

Objectives

To learn the chemical and physical properties of silicon
and the silicates.

Review

Silicon is the second most abundant element on Earth. It
is a metalloid and a semiconductor. Silicon is recovered from
silica by reduction with carbon in an electric furnace. Very
pure silicon is prepared by reduction of $SiCl_4$ with H_2, followed
by zone refining. The element is relatively unreactive, but dissolves in base with the evolution of hydrogen.

The most important compounds of silicon are the silicates.
The principal structural element common to all of them is the
SiO_4 tetrahedron. By sharing corners with other SiO_4 tetrahedra, structures of varying degrees of complexity result.
These can be summarized as follows:

    (a) Simple SiO_4 tetrahedron

$$SiO_4{}^{4-} \qquad \text{(orthosilicate ion)}$$

(b) Sharing of a corner between two SiO_4 tetrahedra

$Si_2O_7^{6-}$ (pyrosilicate ion)

(c) Sharing of two corners by each SiO_4 tetrahedron

$(SiO_3)_n^{2n-}$ (metasilicate ion)

This produces long strands. Double strands are also formed, as shown in Figure 20.14.

(d) Sharing of three corners by each SiO_4 tetrahedron

$(Si_2O_5)_n^{2n-}$

This produces planar sheets.

(e) Sharing all four corners by each SiO_4 tetrahedron

SiO_2 (silica; quartz)

This gives a three-dimensional network of covalent bonds in which the SiO_4 tetrahedra are stacked in coils. Two types of crystals are found, depending on the direction of rotation of these coils.

Silicones are inorganic polymers in which the "backbone" of the polymer is a long chain of alternating silicon and oxygen atoms. They are formed by hydrolysis of compounds such as $(CH_3)_2SiCl_2$, which is accompanied by the elimination of H_2O from pairs of Si—O—H units and the linking together of silicon atoms by oxygen bridges.

Self-Test

36. Write the chemical equation for the production of silicon in an electric furnace.

_____

37. On a separate sheet of paper, draw the Lewis structure of (a) the orthosilicate ion, (b) the pyrosilicate ion, and (c) a portion of a metasilicate ion.

38. Name a mineral that contains

(a) simple SiO_4^{4-} ions   _____

(b) long double-chains of SiO_4 tetrahedra_____

(c) sheets of SiO_4 tetrahedra   _____

(d) the cyclic $Si_6O_{18}^{12-}$ ion   _____

39. What product would you expect from the hydrolysis of $(CH_3)_3SiCl$? Draw its expected structure.

New Terms

zone refining
silicone

Answers to Self-Test Questions

1. $Ca_3(PO_4)_2$

2. $2Ca_3(PO_4)_2 + 6SiO_2 + 10C \longrightarrow 6CaSiO_3 + 10CO + P_4$

3. Compare your answers to the structures shown on Pages 622, 624, and 623.

4. $Na_3P + 3H_2O \longrightarrow 3NaOH + PH_3$

5. (a) $H-\overset{\cdot\cdot}{P}-H$ (b) pyramidal
        $|$
        H

6. (a) $P_4O_6 + 6H_2O \longrightarrow 4H_3PO_3$

   (b) $P_4O_{10} + 6H_2O \longrightarrow 4H_3PO_4$

   (c) $PCl_3 + 3H_2O \longrightarrow H_3PO_3 + 3HCl$

7. A substance that is able to remove water from a gas or liquid.

8. (a) K_2HPO_3 and KH_2PO_3

   (b) By the hydrolysis of PCl_3 (see answer 6c above)

9. $$\left[\begin{array}{ccc} & :\overset{\cdot\cdot}{O}: & :\overset{\cdot\cdot}{O}: \\ & | & | \\ :\overset{\cdot\cdot}{O}-P-\overset{\cdot\cdot}{O}-P-\overset{\cdot\cdot}{O}: \\ & | & | \\ & :\overset{\cdot}{O}: & :\overset{\cdot}{O}: \end{array} \right]^{4-}$$

10. (a) $Na_4P_2O_7$ (b) $Na_5P_3O_{10}$

11. Compare your answers to the structures shown on Page 627.

12. $POCl_3$; the manufacture of flame retardants.

13. The pale yellow nonviscous liquid thickens and darkens as S_8 rings break into S_8 chains that link to form long S_x chains. At still higher temperatures the chains break into smaller pieces and the dark red liquid becomes thin and nonviscous again.

14. $Na_2SO_3 + 2HCl \longrightarrow 2NaCl + H_2O + SO_2$

15. (a) $H_2S_2O_7$ (b) $H_2S_2O_7 + H_2O \longrightarrow 2H_2SO_4$

16.

$$H-\overset{\displaystyle :\overset{..}{O}:}{\underset{\displaystyle :\overset{..}{O}:}{\overset{|}{\underset{|}{S}}}}-\overset{..}{\underset{..}{O}}-H \qquad H-\overset{..}{\underset{..}{O}}-\overset{\displaystyle ..}{\underset{\displaystyle :\overset{..}{O}:}{\overset{|}{\underset{|}{S}}}}-\overset{..}{\underset{..}{O}}-H$$

17. $2I^- + 3H_2SO_4 \longrightarrow I_2 + SO_2 + H_2O + 2HSO_4^-$

18. (a) $ZnS + 2HCl \longrightarrow ZnCl_2 + H_2S$

   (b) $SO_2 + H_2O \longrightarrow H_2SO_3$

   (c) $2S_2O_3^{2-} + AgBr \longrightarrow Ag(S_2O_3)_2^{3-} + Br^-$

   (d) $2S_2O_3^{2-} + I_2 \longrightarrow 2I^- + S_4O_6^{2-}$

   (e) $S + SO_3^{2-} \longrightarrow S_2O_3^{2-}$

19. (a) CaF_2, Na_3AlF_6, $Ca_5(PO_4)_3F$ (b) $NaCl$

20. Br_2, as evidenced by its higher m.p. and b.p.

21. (a) N.R.

   (b) $2FeBr_3 + 3F_2 \longrightarrow 2FeF_3 + 3Br_2$

   (c) $PbI_2 + Cl_2 \longrightarrow PbCl_2 + I_2$

22. $4HCl + MnO_2 \longrightarrow MnCl_2 + Cl_2 + 2H_2O$

23. $2Na^+ + 2Cl^- + 2H_2O \longrightarrow 2Na^+ + 2OH^- + H_2 + Cl_2$

24. $MnO_2 + 2Br^- + 4H^+ \longrightarrow Mn^{2+} + Br_2 + 2H_2O$

25. (a) bromine (b) iodine   26. $TiCl_2$

27. (a) $CaF_2 + H_2SO_4 \longrightarrow CaSO_4 + 2HF(g)$

   (b) $NaCl + H_2SO_4 \longrightarrow NaHSO_4 + HCl(g)$

   (c) $NaBr + H_3PO_4 \xrightarrow{heat} NaH_2PO_4 + HBr(g)$

28. HCl

29. (a) $CaO + Cl_2 \longrightarrow CaCl(OCl)$

   (b) $Cl_2 + 2OH^- \longrightarrow Cl^- + OCl^- + H_2O$

(c) $4KClO_3 \longrightarrow 3KClO_4 + KCl$

(d) $3Br_2 + 6OH^- \longrightarrow BrO_3^- + 5Br^- + 3H_2O$

(e) $SiCl_4 + 2H_2O \longrightarrow SiO_2 + 4HCl$

30. HIO_4

31. See Table 20.2 on Page 640.

32. (a) T-shaped (b) distorted tetrahedral (c) tetrahedral

33. $GeCl_4 + 2H_2O \longrightarrow GeO_2 + 4HCl$

34. The availability of low-energy d orbitals in the valence shell of silicon allows attachment of H_2O as one step in the mechanism for the hydrolysis.

35. (a) nonlinear (AX_2E_2 type molecule)

(b) tetrahedral (AX_4 type molecule)

(c) square planar (AX_4E_2 type molecule)

36. $SiO_2 + 2C \longrightarrow Si + 2CO$

37. (a) see Page 648   (b) and (c) see Page 649

38. (a) zircon (b) asbestos (c) mica, soapstone (d) beryl

39. $(CH_3)_3Si-O-Si(CH_3)_3$

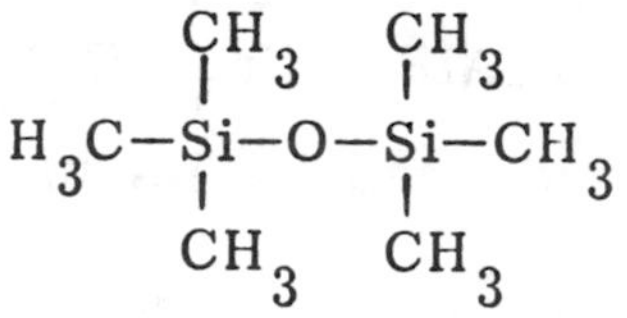

$$
\begin{array}{ccc}
 & CH_3 & CH_3 \\
 & | & | \\
H_3C- & Si-O-Si & -CH_3 \\
 & | & | \\
 & CH_3 & CH_3
\end{array}
$$

21 THE TRANSITION ELEMENTS

This chapter deals with the chemistry of the elements in the center of the periodic table and those in the two rows that are placed just below the main body of the table.

21.1 GENERAL PROPERTIES

Objectives

To examine some general trends and relationships among the physical and chemical properties of the transition elements.

Review

Learn the nomenclature used to refer to the different types and classes of transition elements. Review the properties that most transition metals have in common:

(a) multiple oxidation states
(b) many compounds are paramagnetic
(c) many compounds are colored
(d) strong tendency to form complex ions

New Terms

d-block elements
main transition elements

inner transition elements
triad

21.2   ELECTRONIC STRUCTURE AND OXIDATION STATES

Objectives

To review the electronic structures of the transition elements and to examine trends in the stabilities of oxidation states. You should learn the maximum oxidation states observed in the various B-groups and the relative stabilities of high and low oxidation states.

Review

The electronic structures of the transition elements are given in Tables 21.2 to 21.4. Note the irregularities at Cr and Cu in the first transition series. This is accounted for on the basis of the added stability possessed by a half-filled or filled subshell.

The oxidation states for the d-block elements are summarized in Figure 21.2. There are several points you should know.

(1)   All elements in the first transition series (except Sc) show a 2+ oxidation state.

(2)   For Groups IIIB to VIIB the maximum positive oxidation state is equal to the group number. This also applies to the Group IIB elements and, to some extent, to the Group IB elements.

(3)   In a given period, the higher oxidation states become less stable (relative to the lower ones), moving from left to right.

(4)   In a given group, the higher oxidation states become relatively more stable, moving from top to bottom.

Self-Test

1.   Predict the electron configurations of the following elements:

(a)   Ti   _____   (d)   Co   _____

(b)   Cr   _____   (e)   Cu   _____

(c)   Mn   _____

2. Why do nearly all first transition series elements show a 2+
    oxidation state?
    _____

3. Choose the compound in each pair that is expected to be the
    best oxidizing agent.

    (a) V_2O_5 or Ta_2O_5 _____

    (b) MnO_2 or TiO_2 _____

    (c) Sc_2O_3 or Mn_2O_3 _____

New Terms

21.3   ATOMIC AND IONIC RADII

Objectives

To investigate trends in these two properties.   Learn
what is meant by "lanthanide contraction" and how it affects the chemistry of the elements of the third transition
series.

Review

Only small horizontal changes in size occur among the
transition elements because of the effectiveness of the d electrons
at shielding the outer s electrons from the nucleus.

Large size increases occur from the first to the second
series, but little or no changes occur from the second to the
third series.   This is due to the lanthanide contraction.

Self-Test

4. Molybdenum and tungsten have the same atomic radii, but
    their atoms differ greatly in atomic mass.   The density of
    molybdenum is 10.2 g/cm^3.   Compute the expected density
    of tungsten.
    _____

5. In each pair, which has the higher ionization energy?

(a) Fe or Ru _____   (b) Ru or Os _____

New Terms

21.4   METALLURGY

Objectives

To learn how metals are extracted from their ores and purified to the point of being able to be put to practical use. You should learn the kinds of pretreatments given to ores, the way the metals are extracted from the ores, and the refining methods used to make them useful.

Review

The three principal steps in the commercial production of metals are: concentration, reduction and refining.

Concentration. Pretreatment procedures can be physical or chemical. Physical separations take advantage of differences in physical properties between the metal bearing component of the ore and the unwanted gangue (review flotation, amalgamation). Chemical separations rely on chemical properties to enrich the metal bearing component of the ore (review roasting, purification of Al_2O_3).

Reduction. The methods of extracting metals from their compounds were discussed in Chapter 18. In this chapter, be sure you know the chemistry of the blast furnace - which substances serve as raw materials and which reactions occur in the various stages.

Refining. This involves purification of the metal after reduction as well as the formation of alloys with desirable properties. Review the refining procedures used to produce steel from pig iron.

Self-Test

6. What property of Al_2O_3 is exploited in the purification of bauxite?

7. Write the chemical equation for the roasting of ZnS in air.

_____

8. What is the active reducing agent in the blast furnace?

_____

9. Write the chemical equation for the reaction between CaO and P_2O_5 during the formation of slag in the blast furnace.

_____

10. Why did the open hearth process replace the Bessemer process for the production of steel?

_____

_____

11. Why has the basic oxygen process largely replaced the open hearth process in the production of steel?

_____

_____

New Terms

metallurgy	gangue	pig iron
ore	roasting	cast iron
refining	blast furnace	Bessemer converter
amalgam	charge	open hearth furnace
flotation	slag	basic oxygen process

21.5 MAGNETISM

Objectives

To account for the magnetic properties of the transition metals and their compounds. You should learn the origin of ferromagnetism and how it differs from paramagnetism.

Review

Paramagnetism arises from the presence of unpaired electrons in atoms, molecules, or ions. Ferromagnetism appears to result from the alignment of many paramagnetic ions in domains in the solid state only. Alignment of domains produces a "permanent magnet." The formation of ferromagnetic domains seems to depend on interionic spacings.

<u>Self-Test</u>

12. Which three pure metals are ferromagnetic?

_____

13. Why does melting a ferromagnetic solid produce a paramagnetic liquid?

_____

_____

<u>New Terms</u>

ferromagnetism
domain

21.6   PROPERTIES OF SOME TRANSITION METALS

<u>Objectives</u>

To learn some of the physical and chemical properties of the important transition metals.

<u>Review</u>

<u>Chromium</u>. This hard, brittle, lustrous metal is familiar to everyone. One of its principal uses is in stainless steel. Chromium's major oxidation states are 2+, 3+, and 6+; the most stable with respect to redox is the 3+ state, which exists as $Cr(H_2O)_6^{3+}$ in water. Chromium(III) hydroxide is amphoteric. The CrO_4^{2-} and $Cr_2O_7^{2-}$ ions are strong oxidizing agents. You should know the structures of these ions and the equilibrium that exists between them.

<u>Manganese</u>. This metal is more easily oxidized than chromium. Its major uses are in alloys. The principal oxidation states of Mn are: 2+ (the pale pink $Mn(H_2O)_6^{2+}$ ion), 4+ (MnO_2), 6+ (the green manganate ion, MnO_4^{2-}), and 7+ (deep violet permanganate ion, MnO_4^-). The most stable state is 2+. You should know why MnO_4^- is a useful titrant for redox reactions in acidic solution, and why it is seldom used for titrations in basic solution.

Iron. This is the most used transition metal because of its desirable physical properties and its relative ease of extraction from its compounds. Iron has two principal oxidation states, 2+ and 3+. The 2+ state is generally easily oxidized to the 3+ state. You should know the three oxides of iron and their properties. Study the mechanism believed to be responsible for the rusting of iron.

Cobalt. This metal is useful in alloys. It has two oxidation states, 2+ and 3+. In water, the pink $Co(H_2O)_6^{2+}$ ion is most stable. In the presence of complex ion-forming substances, the 3+ state is often preferred.

Nickel. This metal is very corrosion resistant and is used in stainless steels. Iron-nickel alloys are impact resistant. The most stable oxidation state of nickel is 2+. In water it exists as the emerald green $Ni(H_2O)_6^{2+}$ ion. NiO_2 is important as the cathode material in nickel-cadmium batteries.

The Coinage Metals - copper, silver, and gold. The reactivities of these metals are low, and decrease from copper to silver to gold. They do not dissolve in nonoxidizing acids. Copper and silver dissolve in HNO_3, but gold will only dissolve in aqua regia. Principal oxidation states are: copper (1+, 2+), silver (1+), gold (1+, 3+). Copper(I) disproportionates in water to give Cu and Cu^{2+}. You should learn the qualitative tests for Cu^{2+} and Ag^+.

Zinc, Cadmium, and Mercury. Both Zn and Cd are reactive metals that dissolve in acids such as HCl. Mercury, the only liquid metal at room temperature, does not react with HCl, but does dissolve in HNO_3. Each of these elements has a 2+ oxidation state. Mercury also has a 1+ state where it exists as the ion Hg_2^{2+}. Zinc and cadmium are used to protect steel from corrosion. Coating steel with zinc is called galvanizing. Zinc prevents iron from rusting by providing cathodic protection. Zinc is amphoteric, but cadmium is not, so cadmium is used to protect steel when the environment will be basic. Zinc has a relatively low toxicity, but both cadmium and mercury are quite poisonous. You should learn the qualitative test for mercury(I) salts.

Self-Test

14. Which metals, besides iron, are used to make stainless steel?

_____

15. (a) What equilibrium exists between chromate ion and dichromate ion? _____

    (b) Sketch the structures of these two ions.

16. Write an equation showing how $Cr(H_2O)_6^{3+}$ behaves like a weak acid. _____

17. What is the formula for chromic acid? _____

18. (a) Why is MnO_4^- a useful titrant for redox reactions in acidic solution? _____

    _____

    (b) Why is it less useful when the reaction is to be carried out in basic solution? _____

    _____

19. Under what conditions is the manganate ion stable?

    _____

20. (a) Give the formulas for the three oxides of iron.

    _____

    (b) Which one is magnetic? _____

21. (a) What substances must be present in order for iron to rust? _____

    (b) What is the formula for rust? _____

22. What are the principal oxidation states of

    (a) iron? _____

    (b) cobalt? _____

    (c) chromium? _____

    (d) manganese _____

23. Which of the coinage metals dissolve in HCl? _____

24. When a clear colorless aqueous solution was acidified with HCl, a white precipitate was formed. This was separated from the remainder of the solution and treated with aqueous ammonia, which caused the precipitate to turn black. The solution of ammonia was then separated from the black precipitate and acidified with HNO_3. This caused a white precipitate to form. On the basis of these observations, which metal ions were present in the original aqueous solution?

_____

25. Complete and balance the following equations. If no reaction occurs, write N.R.

   (a) $Mn + HCl \longrightarrow$

   (b) $Fe + H_2SO_4 \longrightarrow$

   (c) $Zn + OH^- \longrightarrow$

   (d) $Cd + OH^- \longrightarrow$

   (e) $Ag + HNO_3 \longrightarrow$

26. What color are the following?

   (a) $Cu(H_2O)_4^{2+}$ _____

   (b) $Mn(H_2O)_6^{2+}$ _____

   (c) $Co(H_2O)_6^{2+}$ _____

   (d) MnO_4^- _____

   (e) $Cu(NH_3)_4^{2+}$ _____

   (f) $HgNH_2Cl$ _____

   (g) $Ni(H_2O)_6^{2+}$ _____

   (h) CrO_4^{2-} _____

   (i) $AgCl$ _____

   (j) $Cr(H_2O)_6^{3+}$ _____

New Terms

rust   galvanizing
coinage metals   cathodic protection

21.7   COORDINATION COMPOUNDS

Objectives

To define what is meant by "coordination compound," to examine the kinds of compounds that are formed, and to review the terminology used to describe them.

Review

Complex ions formed from a metal ion and one or more ligands are called coordination compounds. Ligands are nearly always neutral molecules or negatively charged ions which have a pair of electrons that can be donated in a coordinate covalent bond. The metal and all ligands in the first coordination sphere are generally enclosed within brackets (e.g., $[Co(NH_3)_6]^{3+}$). Monodentate ligands provide one coordinating atom; bidentate ligands provide two and form ring structures. These complexes are called chelates. A polydentate ligand is able to provide more than one donor atom. Important bidentate ligands are:

$$\text{oxalate ion, } C_2O_4^{2-}$$

$$\text{ethylenediamine (en), } H_2N-CH_2-CH_2-NH_2$$

An important polydentate ligand is EDTA.

Self-Test

27. What is the charge on the metal ion in each of the complex ions below?

(a) $[Mn(C_2O_4)_3]^{3-}$ _____

(b) $[FeCl_6]^{4-}$ _____

(c) $[Cr(en)_2Cl_2]^+$ _____

(d) $[Ni(CN)_4]^{2-}$ _____

(e) $[PtCl_6]^{2-}$ _____

New Terms

coordination compound   bidentate ligand
first coordination sphere   polydentate ligand
ligand   chelate
monodentate ligand

21.8   COORDINATION NUMBER

Objectives

To examine the kinds of structures formed when different numbers of atoms are coordinated to the central metal ion.

Review

Learn the definition of coordination number. Review the structures that are found for C.N. = 2, 4 and 6. Learn to draw the 2-dimensional representation of octahedral coordination described in Figure 21.10.

Self-Test

28. Practice drawing, on a separate piece of paper, the geometries observed most commonly for C.N. = 2, 4 and 6.
     Check yourself by referring to Figure 21.9 in the text.

New Terms

coordination number

21.9   NOMENCLATURE

Objectives

To learn how to name coordination compounds. You should be able to write the name, given the formula of a complex; you should be able to write the formula, given the name of a complex.

Review

Learn the nomenclature rules given on Pages 682 and 683 in the text. After you feel you know them, practice on the Self-Test below.

Self-Test

29. Name the following:

    (a) $[Co(NH_3)_6]^{3+}$ _____

    (b) $[CoBr_6]^{3-}$ _____

    (c) $[Mn(en)_2Cl_2]^+$ _____

    (d) $[Ni(H_2O)_6]^{2+}$ _____

    (e) $[Fe(H_2O)_4(NH_3)_2]^{2+}$ _____

    (f) $[Ag(CN)_2]^-$ _____

30. Write the formulas for the following:

    (a) tetraamminedichloronickel(II) _____

    (b) sodium bis(carbonato)dichlorocobaltate(III)

    (c) dithiosulfatoargentate(I) ion _____

    (d) hexaaquachromium(III) hexacyanochromate(III)

    (e) diaquabis(oxalato)chromate(III) ion _____

    (f) diammineaquachlorodithiocyanatomanganate(II) ion

    (g) dicyanobis(ethylenediamine)cobalt(III) chloride

New Terms

21.10   ISOMERISM AND COORDINATION COMPOUNDS

Objectives

> To learn the meaning of the term "isomerism" and to see how it is applied to coordination compounds. You should learn the different types of isomerism described in this section.

Review

Compounds having the same formula but different structures are called isomers. Review the following types of isomerism: ionization isomerism, stereoisomerism, geometrical isomerism, optical isomerism. Be sure you know the difference between cis- and trans- isomers. Remember that optical isomers are nonsuperimposable mirror images of each other and are said to be chiral. Optically active compounds have the ability to rotate plane polarized light. An equal mixture of optical isomers is said to be racemic.

Self-Test

31. Sketch the cis-trans isomers for (a) $[Co(H_2O)_4Cl_2]^+$ and (b) $[Co(C_2O_4)_2Cl_2]$.

32. Which of the following are expected to exhibit optical isomerism?

(a) $[Co(NH_3)_6]^{3+}$

(b) trans-$[Co(C_2O_4)_2Cl_2]^{3-}$

(c) $[Co(en)_3]^{3+}$

(d) cis-$[Co(en)_2(NH_3)_2]^{3+}$

(e) cis-$[Co(NH_3)_4Cl_2]^+$ _____

New Terms

isomer
ionization isomerism
stereoisomerism
geometrical isomerism
optical isomerism
cis-
trans-

chiral
polarized light
enantiomers
optical activity
dextro-rotatory
levo-rotatory
racemic

21.11   BONDING IN COORDINATION COMPOUNDS: VALENCE BOND THEORY

Objectives

To account for the structure and magnetic properties of complex ions using the valence bond theory.

Review

To form coordinate covalent bonds, the metal ion must provide one empty hybrid orbital for each coordinating ligand atom. An inner orbital complex is formed if the d subshell below the metal atom's outer shell is used. For C.N. = 6 the hybrids are specified as d^2sp^3. An outer orbital complex is formed if the d orbitals are from the metal atom's outer shell. For C.N. = 6, these hybrids are specified as sp^3d^2.

When the metal ion has four, five or six d electrons, it is necessary to choose between inner and outer orbital complexes because they lead to different magnetic properties. As a general rule, for d^4 or d^6 ions of the first transition series, inner orbital complexes are preferred except when the ligands are F^- or H_2O. For d^5 ions, outer orbital complexes are preferred except when the ligand is NO_2^- (nitro) or CN^-.

Tetrahedral complexes use sp^3 hybrid orbitals; square planar complexes use dsp^2 hybrids.

Self-Test

33.   Give the orbital diagrams for the following octahedral com-

plex ions.

(a) $[MnCl_6]^{2-}$

(b) $[FeCl_6]^{3-}$

(c) $[Co(CN)_6]^{3-}$

New Terms

inner orbital complex   pairing energy
outer orbital complex

21.12 CRYSTAL FIELD THEORY

Objectives

To describe a bonding theory that can explain the colors of complex ions as well as their magnetic properties.

Review

Crystal field theory considers the effect of the ligand ions (or dipoles) on the energies of the d orbitals of the metal ion. In an octahedral complex the ligands split the d orbitals into a low energy set of three orbitals (the t_{2g} level) and a high energy set of two orbitals (the e_g level).   The energy difference between them is called Δ. When a complex absorbs light, an electron is raised in energy from the t_{2g} to the e_g level. The color of the light absorbed depends on the magnitude of Δ. The size of Δ is influenced by the nature of the ligands. Learn the shapes of the d orbitals (Figure 21.20) and the spectrochemical series given on Page 695.

For metal ions with d^4, d^5, d^6 or d^7 configurations the magnetic properties are determined by the magnitude of Δ in relationship to the pairing energy P. Review the discussion of the factors that determine whether a low spin or high spin complex is formed.

Review the splitting patterns of the d orbitals for tetrahedral and square planar complexes. Remember that Δ_{tetr} is always much less than Δ_{oct}.

Self-Test

34. On a separate sheet of paper, sketch the CFT splitting pattern for d orbitals in octahedral, square planar, and tetrahedral complex ions. Check your answers by referring to Figures 21.27 and 21.29 in the text.

35. On a separate sheet of paper, indicate the electron population of the t_{2g} and e_g orbitals in low and high spin complexes for a d^7 metal ion.

36. Which complex in each pair below should absorb light of highest frequency (shortest wavelength)?

   (a) $[CrCl_6]^{3-}$ or $[CrBr_6]^{3-}$ _____

   (b) $[CrCl_6]^{3-}$ or $[Cr(NH_3)_6]^{3+}$ _____

   (c) $[NiCl_6]^{4-}$ or $[Ni(NO_2)_6]^{4-}$ _____

   (d) $[Fe(NH_3)_6]^{3+}$ or $[Fe(CN)_6]^{3-}$ _____

37. In a certain complex containing a metal ion with a d^6 configuration, the pairing energy is greater than Δ. Will this complex be low spin or high spin?

   _____

New Terms

crystal field theory   low spin complex

spectrochemical series   high spin complex

Answers to Self-Test Questions

1. See Table 21.2 in the text.   2. From the loss of two 4s electrons   3.(a) V_2O_5 (b) MnO_2 (c) Mn_2O_3 4. d = 19.5 g/cm^3 (calculated). The atomic weight of W (183.85) is 1.92 times larger than that of Mo (93.94), so 1.92 times as much mass is packed into the same volume. Therefore, the density of W should be 1.92 times larger than that of Mo. The actual measured density of tungsten is 19.3 g/cm^3. 5.(a) Fe (b) Os 6. Its amphoteric nature.
7. $2ZnS + 3O_2 \longrightarrow 2ZnO + 2SO_2$ 8. CO

9. $3CaO + P_2O_5 \longrightarrow Ca_3(PO_4)_2$ 10. More uniform properties could be obtained.   11. It is much faster.   12. iron, cobalt, and nickel   13. Melting destroys the domains, which are necessary for ferromagnetism.   14. chromium and nickel

15.(a) $CrO_4^{2-} + 2H^+ \rightleftharpoons Cr_2O_7^{2-} + H_2O$ (b) See Page 671.

16. $Cr(H_2O)_6^{3+} + H_2O \rightleftharpoons Cr(H_2O)_5OH^{2+} + H_3O^+$ 17. H_2CrO_4

18.(a) MnO_4^- is deep violet, Mn^{2+} is very pale pink; MnO_4^- serves as its own indicator.

   (b) MnO_4^- gives solid MnO_2 in basic solution, which obscures the endpoint.

19. Very basic conditions

20.(a) FeO, Fe_2O_3, Fe_3O_4 (b) Fe_3O_4, magnetite

21.(a) oxygen <u>and</u> moisture   (b) $Fe_2O_3 \cdot xH_2O$

22.(a) 2+, 3+   (b) 2+, 3+   (c) 2+, 3+, 6+   (d) 2+, 4+, 6+, 7+
23. none of them   24. Hg_2^{2+} and Ag^+

25.(a) $Mn + 2HCl \longrightarrow MnCl_2 + H_2$

   (b) $Fe + H_2SO_4 \longrightarrow FeSO_4 + H_2$

   (c) $Zn + 2OH^- + 2H_2O \longrightarrow Zn(OH)_4^{2-} + H_2$

   (d) N.R.

   (e) $3Ag + 4HNO_3 \longrightarrow 3AgNO_3 + NO + 2H_2O$

26.(a) pale blue   (b) pale pink   (c) pink   (d) deep violet
   (e) deep blue   (f) white   (g) green   (h) yellow   (i) white
   (j) violet   27.(a) +3   (b) +2   (c) +3   (d) +2   (e) +4
28. See Figure 21.9.   29.(a) hexaamminecobalt(III) ion
(b) hexabromocobaltate(III) ion
(c) dichlorobis(ethylenediamine)manganese(III) ion
(d) hexaaquanickel(II) ion   (e) diamminetetraaquairon(II) ion
(f) dicyanoargentate(I) ion
30.(a) $[Ni(NH_3)_4Cl_2]$ (b) $Na_3[Co(CO_3)_2Cl_2]$ (c) $[Ag(S_2O_3)_2]^{3-}$

(d) $[Cr(H_2O)_6][Cr(CN)_6]$ (e) $[Cr(C_2O_4)_2(H_2O)_2]^+$

(f) $[Mn(NH_3)_2(SCN)_2(H_2O)Cl]^-$ (g) $[Co(en)_2(CN)_2]Cl$

31. (a)

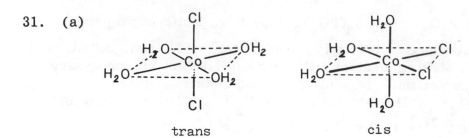

trans   cis

(b)

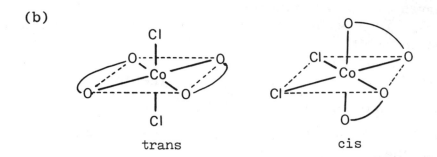

trans   cis

32. c and d

33. (a) $\uparrow$ $\uparrow$ $\uparrow$ XX XX   XX   XX XX XX   __ __ __ __ __
           3d   4s   4p   4d

(b) $\uparrow$ $\uparrow$ $\uparrow$ $\uparrow$ $\uparrow$ XX   XX XX XX   XX XX __ __ __

(c) $\uparrow\downarrow$ $\uparrow\downarrow$ $\uparrow\downarrow$ XX XX   XX   XX XX XX   __ __ __ __ __

34. See Figures 21.27 and 21.29.

35. $\uparrow$ $\uparrow$ e_g $\uparrow$ __

    $\uparrow\downarrow$ $\uparrow\downarrow$ $\uparrow$ t_{2g} $\uparrow\downarrow$ $\uparrow\downarrow$ $\uparrow\downarrow$

    high spin   low spin

36. (a) $[CrCl_6]^{3-}$ (b) $[Cr(NH_3)_6]^{3+}$ (c) $[Ni(NO_2)_6]^{3-}$
    (d) $[Fe(CN)_6]^{3-}$

37. high spin

22 ORGANIC CHEMISTRY

In this chapter we deal with the chemistry of carbon compounds. These range from simple molecules such as methane, CH_4, to extremely large polymer molecules such as polystyrene or polyethylene. The purpose of the chapter is to acquaint you with the breadth of the subject, rather than to delve too deeply into specific areas, and to illustrate some of the many practical uses to which organic compounds are applied.

22.1 HYDROCARBONS

Objectives

To examine the class of organic compounds called hydrocarbons. You should learn the nomenclature for the first ten members of the alkane, alkene and alkyne series. You should also become familiar with the bonding and structure of these compounds.

Review

Saturated hydrocarbons (the alkanes) contain only single bonds; unsaturated hydrocarbons contain either carbon-carbon double bonds (alkenes) or triple bonds (alkynes). In the alkanes the carbon is sp^3 hybridized and lies at the center of a tetrahedron. When doubly bonded, carbon employs sp^2 hybrid orbitals; when triply bonded, carbon utilizes sp hybrids.

Review the names of the alkanes in Table 22.1. You

should be able to identify the number of carbon atoms in the
chain by the stem of the name; for example, <u>pentane</u> signifies a
<u>five</u> carbon atom chain.

Self-Test

1. Indicate the number of carbon atoms in each of the following:

    (a) butene _____   (e) ethene _____

    (b) hexyne _____   (f) nonene _____

    (c) octane _____   (g) propyne _____

    (d) methane _____   (h) heptene _____

2. Write the molecular formulas for the following:

    (a) propene _____   (c) decyne _____

    (b) butane _____   (d) pentane _____

3. How many π-bonds would be found in:

    (a) hexene _____   (d) C_5H_8 _____

    (b) butylene _____   (e) C_8H_{18} _____

    (c) C_6H_{12} _____

New Terms

saturated hydrocarbon   olefin
unsaturated hydrocarbon   paraffin
alkane   hydrocarbon
alkene   aliphatic hydrocarbons
alkyne   aromatic hydrocarbons
homologous series

22.2   ISOMERS IN ORGANIC CHEMISTRY

Objectives

To examine the types of isomerism that occur among organic compounds.

Review

Structural isomers occur with the alkanes by branching of chains. Among alkenes and alkynes there are also different possible positions of the double or triple bond. Stereoisomerism includes cis-trans isomerism (geometrical isomers) and optical isomerism. Cis-trans isomers occur in alkenes, where free rotation about the carbon-carbon double bond cannot occur. Optical isomers exist when a molecule contains one or more asymmetric carbon atoms (carbon atoms attached to four different groups).

Self-Test

4. On a separate sheet of paper draw the cis and trans isomers of butene. Check your answer by turning to Page 709 of the text.

5. Which of the following can exist as optical isomers?

(a)

$$\begin{array}{c} H \\ | \\ Cl-C-CH_3 \\ | \\ F \end{array}$$

(b)

$$\begin{array}{c} CH_3-CH-CH_3 \\ | \\ CH_3 \end{array}$$

(c) $CH_3-CH_2-CH-CH_2-CH_2-CH_3$
        $\quad\quad\quad\quad\ |$
        $\quad\quad\quad\quad CH_3$

New Terms

structural isomers   geometrical isomers
asymmetric carbon atom

22.3   NOMENCLATURE

Objectives

To describe the nomenclature system that has been devised to name organic compounds. You should be able to name straight and branched-chain alkanes, alkenes, and alkynes.

Review

Review the nomenclature rules and examples on Pages 711 to 716 of the text. When you feel confident that you have learned the rules, try the Self-Test for this section.

Self-Test

6. Name the compounds in Table 22.2 in your text.

7. Name the following:

(a) 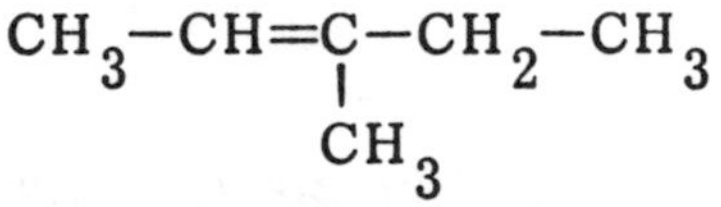 $CH_3-CH=C-CH_2-CH_3$
                    |
                   CH_3

_____

(b) 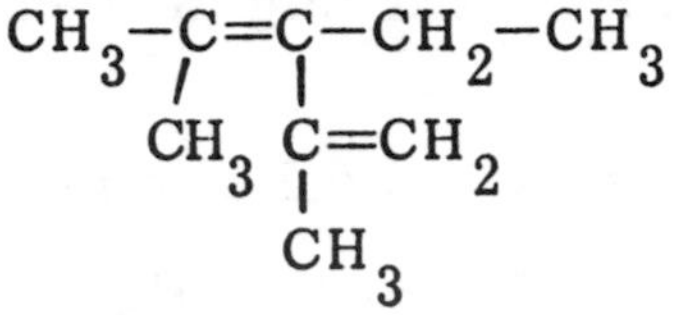 $CH_3-C=C-CH_2-CH_3$
              / |
           CH_3 $C=CH_2$
                   |
                  CH_3

_____

New Terms

alkyl group

22.4   CYCLIC HYDROCARBONS

Objectives

To examine a class of hydrocarbons composed of ring structures.

Review

Learn the nomenclature that is applied to these cyclic compounds. Note that their stability and structure can be related to the C—C—C angle within the ring structure.

Self-Test

8. Write structural formulas for the cyclic hydrocarbons represented by

(a)

_____

(b)

_____

(c)

_____ ___

9. Name the compounds in Question 8.

   (a) _____

   (b) _____

   (c) _____

New Terms

cyclo-

22.5   AROMATIC HYDROCARBONS

Objectives

   To study the structure, bonding and nomenclature of benzene related compounds.

Review

   Remember that benzene has a planar ring structure that can be viewed as a resonance hybrid of two structures with alternating single and double bonds. Molecular orbital theory views the bonding in terms of a delocalized π-electron cloud.

   Review the nomenclature system for benzene derivatives.

Self-Test

10. What is the C—C—C bond angle in benzene?_____

11. Name the following:

(a)

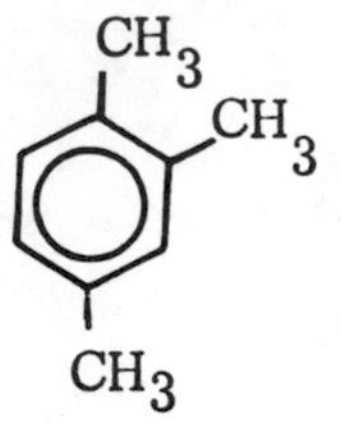

(b)

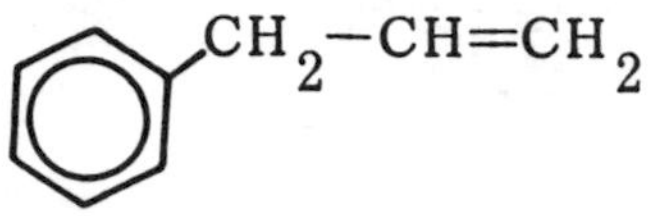

(c)

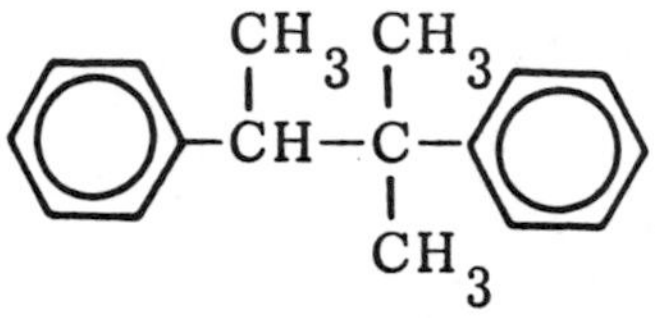

New Terms

ortho   para
meta   phenyl group

22.6   HYDROCARBON DERIVATIVES

Objectives

> To see how most organic compounds can be considered to
> be derived from hydrocarbons by substituting certain
> groups of atoms (called functional groups) for hydrogen
> in the parent hydrocarbon molecule.

Review

A functional group is some group of atoms that imparts a
characteristic property to an organic molecule. Review the functional groups in Table 22.5 on Page 725 of the text.

Typical reactions discussed in this section include addition
reactions, which are characteristic of unsaturated hydrocarbons.
Markovnikov's rule states that when HX is added across a double
bond, the H goes to the carbon atom already containing the most
H's. Remember that saturated hydrocarbons primarily undergo
substitution reactions.

Self-Test

12. Without referring to Table 22.5, identify the following functional groups:

(a)

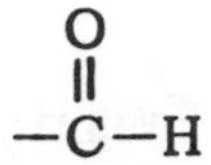

$$-\overset{\displaystyle O}{\overset{\displaystyle \|}{C}}-H$$

_____

(b) −OH

_____

(c)

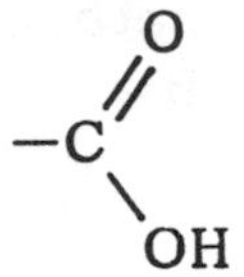

$$-C\overset{\displaystyle O}{\underset{\displaystyle OH}{}}$$

_____

(d)

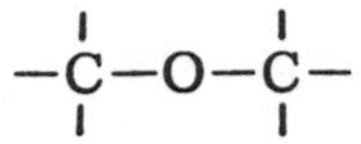

$$-\overset{|}{C}-O-\overset{|}{C}-$$

_____

(e)

$$-\overset{\displaystyle O}{\overset{\displaystyle \|}{C}}-O-\overset{|}{C}-$$

_____

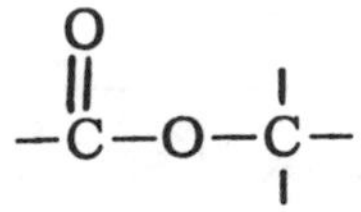

13. Write chemical equations for the following, showing the structures of the main product:

(a) $CH_3-CH=CH_2 + Br_2 \longrightarrow$ _____

(b) $CH_3-CH_3 + Cl_2 \longrightarrow$ _____

(c) $CH_3-\underset{\displaystyle \underset{|}{CH_3}}{C}=CH_2 + HCN \longrightarrow$

(This is an addition reaction, too.)   _____

New Terms

functional group   substitution reaction
addition reaction   Markovnikov's rule

22.7   HALOGEN DERIVATIVES

Objectives

To learn about some methods of preparation and characteristic reactions of halogenated hydrocarbons. You should also become aware of some of the many uses of these compounds.

Review

Many common chemicals contain halogens. Examine some of these examples on Page 727. Halogenated compounds can be prepared by substitution reactions of halogens with alkanes, and by halogen addition to alkenes. The most important methods involve reactions in which the OH group of an alcohol is displaced by a halogen.

Self-Test

14. List some uses of halogen-containing compounds. _____

    _____

    _____

15. Give an example of a reaction that might be used to prepare $CH_3-CHCl-CH_3$ from $CH_3-CHOH-CH_3$.

    _____

New Terms

22.8   IMPORTANT OXYGEN-CONTAINING DERIVATIVES

Objectives

To examine some properties and chemical reactions involving a number of interrelated oxygen-containing functional groups. You should also learn some applications of the various kinds of compounds discussed in this section.

Review

Six major functional groups are discussed in this section: alcohols, aldehydes, ketones, acids, esters, and ethers. Be sure you can distinguish among them.

Some important relationships in this section are:

(1) Alcohols can be oxidized to give aldehydes or ketones; aldehydes can be oxidized to give acids.

(2) The products of oxidation of an alcohol depend on whether the alcohol is primary, secondary, or tertiary.

(3) Acids react (reversibly) with alcohols to produce esters by a condensation reaction involving elimination of water. The base-catalyzed hydrolysis of an ester is called saponification.

(4) Aldehydes and ketones contain the carbonyl group, $>C=O$.

(5) The acidity of organic acids arises from the carboxyl group, $-COOH$.

(6) Condensation of two alcohols produces an ether.

Self-Test

16. Give some uses for:

(a) alcohols _____

(b) ketones _____

(c) esters _____

(d) ethers _____

17. Give the major organic product in each of the following:

(a)
$$CH_3-\overset{\overset{\textstyle O}{\|}}{C}-O-C_2H_5 + H_2O \xrightarrow{H^+}$$ _____

(b) $CH_3-\underset{\underset{\textstyle OH}{|}}{CH}-CH_3 \xrightarrow{\text{oxidation}}$ _____

(c) $CH_3OH + CH_3COOH \longrightarrow$ _____

(d)
$$CH_3-\overset{\overset{\textstyle O}{\|}}{C}-CH_3 + NaBH_4 \longrightarrow$$ _____

New Terms

alcohol

carbinol group

primary alcohol

secondary alcohol

tertiary alcohol

aldehyde

ketone

carbonyl group

organic acid

carboxyl group

ester

esterification

condensation reaction

saponification

ether

22.9   AMINES AND AMIDES

Objectives

To examine this class of organic compounds that can be considered to be derivatives of ammonia.

Review

Remember the functional groups for amines and amides. Learn to distinguish between primary, secondary, and tertiary amines. Amines give basic aqueous solutions for the same reasons that solutions of ammonia are basic (the lone pair of electrons on the nitrogen atom of the amine is capable of accepting a proton from water).

Self-Test

18. Draw structural formulas for

    (a)   dimethyl amine   (c)   urea

    (b)   pyridine

19. Write a chemical equation showing how dimethyl amine behaves as a base in water.

New Terms

amine   heterocycle
amide

22.10 POLYMERS

Objectives

   To learn the general types of polymers that are formed
   and how their properties are related to their structure.

Review

   Polymers are built by linking many small units (monomers)
together to give long chains. Two polymer types exist. Addition
polymers are formed by adding one monomer unit to another.
Polyvinyl chloride is an example (you've probably read that vinyl
chloride monomer has been linked to cancer). Condensation
polymers are formed by elimination of a small molecule (e.g.,
H_2O) from between two monomer units. The polymeric oxoacids
in Chapter 20 are condensation polymers.

   Structural strength in polymers is improved by cross-linking between polymer chains (e.g., in Bakelite and rubber).

Self-Test

20. Which of the following polymers are addition polymers:
     polystyrene, dacron, nylon, polyvinyl chloride, teflon?

     _____

New Terms

monomer   polyester
addition polymer   cross-linking
condensation polymer   vulcanization
copolymer

Answers to Self-Test Questions

1.(a) 4 (b) 6 (c) 8 (d) 1 (e) 2 (f) 9 (g) 3 (h) 7
2.(a)C_3H_6 (b) C_4H_{10} (c) $C_{10}H_{18}$ (d) C_5H_{12} 3.(a) one
(b) one (c) one (d) two (e) none 4. See Page 709.
5. optical isomers found for (a) and (c) 6.(a) hexane
(b) 2-methylpentane (c) 3-methylpentane
(d) 2,3-dimethylbutane (e) 2,2-dimethylbutane
7.(a) 3-methyl-2-pentene (b) 2,4-dimethyl-3-ethyl-2,4-pentadiene
8.(a)

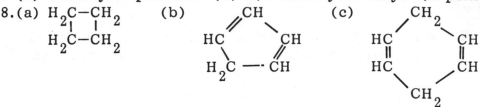

9.(a) cyclobutane (b) 1,3-cyclopentadiene
(c) 1,4-cyclohexadiene 10. 120° 11.(a) 1,2,4-trimethylbenzene
(b) 3-phenyl-1-propene (c) 2-methyl-2,3-diphenylbutane (Did
you name it as a derivative of ethane?) 12. Check your answer
in Table 22.5. 13.(a) $CH_3-CHBr-CH_2Br$ (b) $CH_3CH_2Cl + HCl$
plus other substituted ethanes (c) $CH_3-C(CH_3)CN-CH_3$
14. See Page 727. 15. $CH_3-CHOH-CH_3 + HCl \longrightarrow$
$CH_3-CHCl-CH_3 + H_2O$ 16.(a) See Pages 728-730. (b) See
Pages 731-732. (c) See Pages 732-734. (d) See Page 735.
17.(a) $CH_3COOH + C_2H_5OH$ (b) $CH_3-CO-CH_3$
(c) $CH_3-COO-CH_3$ (d) $CH_3-CHOH-CH_3$
18.(a)

$$H-\underset{\underset{CH_3}{|}}{\overset{\overset{CH_3}{|}}{N}}$$

(b)

(c)

$$H_2N-\overset{\overset{\text{O}}{\|}}{C}-NH_2$$

19. $(CH_3)_2NH + H_2O \rightleftharpoons (CH_3)_2NH_2^+ + OH^-$

20. polystyrene, polyvinyl chloride, teflon

23 BIOCHEMISTRY

Biochemistry is the study of the chemical reactions that take place in living systems, and it is one of the most important areas of chemical research today. The remarkable breakthroughs that have occurred over the past 20 years or so have produced an impressive list of Nobel Prize winners. In this chapter we examine four important kinds of biomolecules. Read the chapter with an eye toward understanding how these substances are usually composed of simple building blocks and how the structures of biomolecules control their properties and biochemical activity.

23.1 PROTEINS

Objectives

To learn how proteins are constructed from amino acids and how protein molecules twist and bend to assume shapes that control their biological functions.

Review

The important ideas developed in this section are the following:

(1) The nature of an α-amino acid. What are the main features of an α-amino acid?

(2) The peptide bond. What is it? How is it formed?

(3) The primary structure of a protein. This is the sequence of amino acids in the peptide chain.

(4) The secondary structure of a protein. The α-helix is an example. How is the structure held in place?

(5) The tertiary structure of a protein. This is found in globular proteins and is controlled by hydrogen bonding, ionic interactions, interactions of nonpolar groups with the polar solvent (water), and the formation of disulfide bridges between cysteine molecules at different places along the chain.

(6) The quaternary structure of some proteins. This concerns the packing of globular proteins into more complex structures.

In addition, the structure of heme is described. The main thing to remember is that the heme group contains an iron atom held in a square planar ligand called a porphyrin. A similar structure is found for chlorophyll.

<u>Self-Test</u> (Use a separate sheet of paper to answer the following three questions.)

1. Sketch the general structure of an α-amino acid.

2. Draw the backbone for the primary structure of a tripeptide.

3. What is a disulfide bridge?

<u>New Terms</u>

protein	secondary structure
α-amino acid	tertiary structure
peptide	quaternary structure
peptide bond	α-helix
peptide linkage	disulfide bridge
polypeptide	porphyrin
primary structure	heme group

23.2 ENZYMES

<u>Objectives</u>

To see how proteins serve as catalysts to promote very

specific biochemical reactions.

Review

The most important idea developed here is the "lock and key" relationship between the enzyme and the molecule upon which it acts (the substrate). This is illustrated for chymotrypsin, but holds for other enzyme-substrate interactions as well.

Review the mechanism of enzyme inhibition.

Self-Test

4. How is an enzyme irreversibly poisoned?

_____

5. What is competitive inhibition?

_____

New Terms

enzyme   substrate
coenzyme   inhibition

23.3 CARBOHYDRATES

Objectives

To examine the structure and properties of the class of compounds called carbohydrates.

Review

These substances are called carbohydrates because their formulas are often of the form, $C_n(H_2O)_m$. They are in fact condensation polymers of monosaccharides (polyhydroxy alcohols also containing an aldehyde or ketone functional group). You should know that the monosaccharides usually exist in cyclic structures. Glucose, one of the most important monosaccharides, can exist in two forms (α-D-glucose and β-D-glucose).

Sucrose is a disaccharide. You should know the general features of the glycoside linkage.

Starch and cellulose are polymers of glucose units. Starch is formed from α-D-glucose units; cellulose from β-D-glucose units.

Self-Test

6. What are the molecular formulas for the following?

    (a) sucrose _____   (b) glucose _____

7. What is the empirical formula for amylose (starch)?_____

8. On a separate sheet of paper, sketch the formation of a glycoside linkage. Compare your answer to Figure 23.15 on Page 756.

9. What characterizes the names of carbohydrates?

_____

New Terms

carbohydrate   polysaccharide
monosaccharide   glycoside linkage
simple sugar

23.4 LIPIDS

Objectives

To study the nature of this class of compounds and to see how lipids are employed by organisms.

Review

Lipids are water insoluble substances found in fats and in cell membranes. Neutral lipids are esters of glycerol and fatty acids (long hydrocarbon chains with a carboxyl group on one end). Saponification of a triglyceride gives glycerol and the anions of the fatty acids. The latter constitute a soap. Review micelle formation in solutions of soap.

Phospholipids contain two fatty acid molecules plus phosphoric acid esterified to glycerol. The phosphoric acid is also esterified to another alcohol. Phospholipids are an important component of cell membranes, forming a bilayer structure.

Another class of lipids are steroids which possess a fused ring system and display very high biological activity.

Self-Test (Use a separate sheet of paper to answer the next four questions.)

10. Draw the general structure of a nonpolar lipid. Write a reaction to show what happens when the lipid molecule is saponified. Check your answer on Pages 758 and 759 of the text.

11. Sketch a micelle formed from anions of fatty acids. You can check your answer on Page 760 of the text.

12. Sketch a bilayer structure typically formed by phospholipids. Check your answer on Page 761.

13. Draw the fused ring structure characteristic of steroids. Check your answer on Page 761.

New Terms

lipid   micelle
fatty acid   phospholipid
triglyceride   bilayer structure
soap   steroid

23.5 NUCLEIC ACIDS

Objectives

To study the composition of nucleic acids (DNA and RNA) and to see how the double helix structure of DNA can account for the transmission of genetic information from one generation to another.

Review

Key points to learn from this section are that nucleic acids are composed of three parts: a nitrogenous base, a five-carbon sugar (ribose or deoxyribose), and phosphoric acid. You should familiarize yourself with the structures of ribose and deoxyribose.

A nucleoside is formed from the nitrogenous base and the sugar. Addition of the phosphoric acid gives a nucleotide - the monomer unit in the DNA or RNA chain. Nucleotides are linked to give the nucleic acid.

In DNA two nucleic acid strands intertwine to give the double helix. These are matched and held together by base-pairing through hydrogen bond formation. Replication involves untwining the two strands and building new complementary strands against each of them, in which the original nuclei acid strands serve as templates.

Self-Test (Use a separate sheet of paper to answer the next two questions. Check your answers by referring to the text.)

14. Sketch the structure of ribose and deoxyribose.

15. Sketch the structure of (a) a nucleoside; (b) a nucleotide.

16. Which bases are able to pair in DNA? _____

_____

New Terms

DNA   nucleic acid
RNA   ribose
nucleotide   deoxyribose
nucleoside   double helix

23.6 PROTEIN SYNTHESIS

Objectives

To learn how DNA in the nucleus of a cell serves to provide the key to determining the primary structure of proteins.

Review

Study the functions of DNA, mRNA, and tRNA as well as the pairing of bases between DNA and RNA. Notice that uracil is found in RNA instead of thymine.

Self-Test

17. Use the DNA/mRNA base pairing scheme to deduce the base sequence that would occur in mRNA if the following base sequence occurred in DNA.

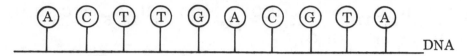

_____DNA

_____

18. Use the genetic code in Table 23.3 to identify the amino acid sequence that could be constructed from mRNA with the following base sequence.

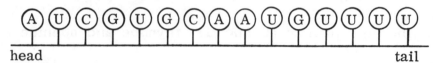

head   tail

_____

19. What amino acid sequence would occur if the fourth base were removed from the mRNA in Question 18?

_____

New Terms

messenger RNA   genetic code
transfer RNA   genetic disease
codon

Answers to Self-Test Questions

1. $R-\underset{\underset{NH_2}{|}}{CH}-COOH$

2. $H_2N-\underset{\underset{R_1}{|}}{CH}-\underset{\underset{O}{||}}{C}-NH-\underset{\underset{R_2}{|}}{CH}-\underset{\underset{O}{||}}{C}-NH-\underset{\underset{R_3}{|}}{CH}-\underset{\underset{O}{||}}{C}-OH$

3. $R_1-S-S-R_2$ where R_1 and R_2 belong to different parts of

the protein backbone

4. When the inhibitor becomes permanently bound to the enzyme active site, the enzyme molecule becomes inoperative.

5. There is a competition between the substrate and inhibitor for the active site on the enzyme.

6.(a) $C_{12}H_{22}O_{11}$, or $C_{12}(H_2O)_{11}$ (b) $C_6H_{12}O_6$ or $C_6(H_2O)_6$

7. $C_6H_{10}O_5$ or $C_6(H_2O)_5$

8. See Figure 23.15.

9. They end in ose (e.g., sucrose).

10. See Pages 758 and 759.

11. See Page 760.

12. See Page 761.

13. See Page 761.

14. See Page 756.

15. See Page 763.

16. cytosine and quanine (C and G); thymine and adenine (T and A)

17.
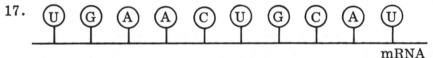

mRNA

18. ile-val-gln-cys-phe

19. ile-cys-asn-val

24 NUCLEAR CHEMISTRY

In this final chapter we will examine nuclear changes, what their implications are in chemistry, and how they can be harnessed in energy production. This latter topic, of course, has become an important issue lately because of growing demands for energy and the increasing price of energy-rich fuels.

24.1 SPONTANEOUS RADIOACTIVE DECAY

Objectives

To review the types of nuclear decay processes that occur and to learn how to write nuclear equations.

Review

You should be sure you know the mass, charge and symbol for each of the particles in Table 24.1. If necessary, review the method of indicating mass number and atomic number for isotopes in Section 3.7, Page 68 of the text.

In balancing a nuclear equation, remember that mass and charge must both be balanced. This requires that the sum of mass numbers (left superscript) on both sides of the arrow must be equal. Also, the algebraic sum of charges (left subscripts) on both sides of the arrow must be the same. Review Example 24.1 in the text.

Kinetically, radioactive decay is a first-order process. The relative rate of decay of radioisotopes are usually expressed in terms of the half-life, $t_{\frac{1}{2}}$, which is the time required for half of the sample to decay. Learn to apply Equations 24.1 and 24.2 (review Examples 24.2 and 24.3).

Archaeological dating is described as an application of radioactive decay processes. Review Example 24.4.

Self-Test

1. Without referring to the text, write the symbols for the following. Afterward, check your answer by referring to Table 24.1 in the text.

   (a) proton _____   (d) alpha particle _____

   (b) neutron _____   (e) positron _____

   (c) beta particle _____

2. Fill in the missing symbol in each of the following:

   (a) $^{27}_{13}\text{Al} + ^{4}_{2}\text{He} \longrightarrow$ _____ $+ ^{1}_{0}\text{n}$

   (b) $^{63}_{29}\text{Cu} + ^{1}_{1}\text{H} \longrightarrow$

       _____

   (c) $^{133}_{56}\text{Ba} + ^{0}_{-1}\text{e} \longrightarrow$

       _____

   (d) $^{213}_{83}\text{Bi} \longrightarrow ^{209}_{81}\text{Tl} +$ _____

   (e) $^{209}_{82}\text{Pb} \longrightarrow ^{209}_{83}\text{Bi} +$ _____

   (f) $^{238}_{92}\text{U} +$ _____ $\longrightarrow ^{247}_{99}\text{Es} + 5\,^{1}_{0}\text{n}$

   (g) $^{235}_{92}\text{U} + ^{1}_{0}\text{n} \longrightarrow ^{139}_{36}\text{Ba} +$ _____ $+ 3\,^{1}_{0}\text{n}$

3. The half-life of $^{90}_{38}\text{Sr}$ is 19.9 years. What is the rate constant for the decay process?

   _____

4. A piece of wood taken from a burial mound of ancient civilization was found to contain only 1/3 as much carbon-14 as in a live tree. Estimate the age of the wood (and the burial mound).

   _____

New Terms

nuclide   radioactive series
radionuclide   decay series
radioisotope   Geiger–Müller counter
parent and daughter isotopes

24.2   NUCLEAR TRANSFORMATIONS

Objectives

To briefly examine how nuclear reactions can be brought
about by bombarding nuclei with various particles.

Review

Nuclear transformations occur when target nuclei are bombarded with various particles.   Learn the shorthand notation used
to describe these reactions (e.g., $^{27}_{13}\text{Al}\,(\alpha,n)\,^{30}_{15}\text{P}$).

New Terms

nuclear transformation   cyclotron
particle accelerator

24.3   NUCLEAR STABILITY

Objectives

To look at the characteristics exhibited by both stable and
unstable nuclei.   You should learn the kinds of reactions
unstable nuclei undergo to become stable.

Review

Stable nuclei lie in a "band of stability." With the exception of hydrogen, stable nuclei always possess at least as many
neutrons as protons; at high atomic number, neutrons outnumber
protons.   The band of stability ends at $Z = 83$.

Elements having an n/p ratio that is too high generally decay by beta emission or neutron emission. Elements with low n/p ratios tend to emit positrons or undergo electron capture. Elements with Z > 83 undergo alpha emission or nuclear fission.

Stable nuclei with odd numbers of both protons and neutrons are rare. Conversely, stable nuclei with even numbers of protons and neutrons are much more common. Some nuclei are very stable and possess "magic numbers" of both protons and neutrons. This supports the nuclear shell theory.

Self-Test

5. The isotope $^{107}_{48}$Cd decays by electron capture. Write the nuclear equation for the reaction.

_____

6. The isotope $^{107}_{49}$In lies below the band of stability. Which of the following is a likely decay mode for this isotope?

(a) $^{107}_{49}$In $\longrightarrow$ $^{107}_{50}$Sn + $^{0}_{-1}$e   (c) $^{107}_{49}$In $\longrightarrow$ $^{103}_{47}$Ag + $^{4}_{2}$He

(b) $^{107}_{49}$In $\longrightarrow$ $^{106}_{49}$In + $^{1}_{0}$n   (d) $^{107}_{49}$In $\longrightarrow$ $^{107}_{48}$Cd + $^{0}_{1}$e

_____

New Terms

band of stability   fission
electron capture   magic number
K capture

24.4 EXTENSION OF THE PERIODIC TABLE

Objectives

To consider the possibility of stable superheavy elements beyond the current range of the periodic table.

Review

The possibility of very heavy stable nuclei is predicted on

the basis of magic numbers of 114 for protons and 184 for neutrons.

<u>Self-Test</u>

7. How many electrons could be accommodated in a g subshell?

    _____

8. Which element (currently known) would be expected to have chemical properties most similar to element Z = 114?

    _____

9. If element 115 were isolated, what would you expect the formula of its oxide to be?

    _____

<u>New Terms</u>

24.5 CHEMICAL APPLICATIONS

<u>Objectives</u>

To examine some chemical applications of radioactive isotopes.

<u>Review</u>

Chemical applications of radioisotopes rely primarily on their ability to be detected and counted. This section illustrates several examples of applications to chemical analysis, the study of descriptive chemistry, and the study of reaction mechanisms.

Review the reasoning involved in the isotope dilution method. Apply this reasoning to Question 10 in the Self-Test. Also review the principles of neutron activation analysis.

<u>Self-Test</u>

10. A 1.00-g portion of KCl labeled with ^{40}K and having a specific activity of 100 cpm per gram was added to a 112-g sample of a mixture of KCl and other salts. By fractional crystallization, some pure KCl was recovered from the mix-

ture. This KCl had a specific activity of 1.2 cpm per gram.
What weight of KCl was in the original 112-g sample? What
percent of the sample was KCl?

_____

11. A student proposed the following structure for the thiosul-
     fate, $S_2O_3^{2-}$, ion: $[O-S-O-S-O]^{2-}$. Thiosulfate is formed
     from SO_3^{2-} by reaction with elemental sulfur.

$$S + SO_3^{2-} \longrightarrow S_2O_3^{2-}$$

If radioactive sulfur is used in this reaction, the $S_2O_3^{2-}$ becomes labeled. When treated with H^+, the $S_2O_3^{2-}$ decomposes and SO_2 is evolved. None of the radioactivity occurs
in this SO_2. How does this argue <u>against</u> the structure
proposed by the student?

New Terms

tracer study   specific activity
isotope dilution   neutron activation analysis

24.6   NUCLEAR FISSION AND FUSION

Objectives

To examine nuclear fission and fusion processes and to
consider them as potential sources of energy.

Review

Fission chain reactions occur because more neutrons are
produced during fission than are consumed. The minimum amount
of fissionable material required to sustain a chain reaction is
called the critical mass.

Review the operation of a nuclear reactor. A breeder
reactor produces more fuel than it consumes.

Nuclear fusion involves the creation of a heavier nucleus
from two lighter ones. Fusion liberates a large amount of energy
but requires extremely high temperatures to occur.

New Terms

fission
fusion
critical mass

breeder reactor
plasma

24.7   NUCLEAR BINDING ENERGY

Objectives

To investigate the source of the large energy changes that occur during fission and fusion.

Review

The mass of a given nucleus is always <u>less</u> than the sum of the masses of the individual protons and neutrons that go toward forming the nucleus.   The difference between the calculated and actual masses is called the mass defect.   Its energy equivalent is called the binding energy and can be calculated from the mass defect by applying Einstein's equation, $E = mc^2$. In these calculations, illustrated on Page 793, use the value,

$$c = 2.99792 \times 10^8 \text{ m/s}$$

The binding energy can be expressed in MeV, in which case,

$$1 \text{ amu} = 931 \text{ MeV}$$

The highest binding energy per nucleon occurs in the vicinity of iron in the periodic table.   Fission releases energy because the lighter particles produced have greater binding energy.   Fusion releases energy because the heavier particles formed have much larger binding energies.

Self-Test

12.   (a)   Compute the binding energy for $^{32}_{16}S$ which has an atomic mass of 31.97207 amu.

_____

(b)   What is the binding energy per nucleon for $^{32}_{16}S$?

_____

New Terms

mass defect   MeV
binding energy

Answers to Self-Test Questions

1. See Table 24.1.
2. (a) $^{30}_{15}P$ (b) $^{64}_{30}Zn$ (c) $^{133}_{55}Cs$ (d) $^{4}_{2}He$ (e) $^{0}_{-1}e$ (f) $^{14}_{7}N$

   (g) $^{139}_{36}Kr$
3. $k = 3.48 \times 10^{-2}\ yr^{-1}$
4. 9000 years
5. $^{107}_{48}Cd + ^{0}_{-1}e \longrightarrow ^{107}_{47}Ag$
6. d
7. 18
8. Pb
9. X_2O_3
10. 83.3 g KCl, 74.4%
11. Since either bond 1 or 2 in O—S—$\overset{1}{O}$—$\overset{2}{}$S—O would be expected to break with equal ease, there is no reason to expect that this structure would never give the labeled sulfur in the SO_2. An unsymmetrical structure is required to explain the results of this experiment. The actual structure of $S_2O_3^{2-}$ is

$$\left[\begin{array}{c} O \\ \| \\ S-S-O \\ \| \\ O \end{array} \right]^{2-}$$

12. (a) 272 MeV (b) 8.50 MeV/nucleon

GLOSSARY

Absolute temperature: The temperature measured on the Kelvin scale.

Absolute zero: $0 \text{ K} = -273.15°C$

Accuracy: How closely an experimental observation lies to the true value.

Acid: In aqueous solutions, a substance that gives H_3O^+. Under the Brønsted-Lowry definition, a proton donor. Under the Lewis definition, an electron-pair acceptor. Under the solvent-system definition, a substance that yields the cation produced in the autoionization of a solvent.

Acid anhydride: A nonmetal oxide that gives an acid when it reacts with water.

Acid dissociation constant: See Acid ionization constant.

Acid ionization constant, K_a: The equilibrium constant for the ionization of a weak acid.

Acid rain: Rain made acidic by dissolved sulfur oxides and nitrogen oxide pollutants.

Acid salt: A salt of a partially neutralized polyprotic acid; for example, $NaHSO_4$ and $NaHCO_3$.

Actinide elements: Elements 90 through 103.

Activated complex: The chemical species that exists with partly broken and partly formed bonds in the transition state.

Activation energy: The minimum kinetic energy that must be possessed by reactant molecules in order to give an effective collision (one that produces the products).

Actual yield: The actual amount of products obtained in a particular chemical reaction when the experiment is performed in the laboratory.

Addition compound: A compound formed by the joining of two molecules with a coordinate covalent bond.

Addition polymer: A polymer formed simply by the joining together, or addition, of monomer units.

Addition reaction: A reaction in which a molecule such as H_2 is added across a double or triple bond in an organic molecule.

Adiabatic: A change that occurs without energy transfer between the system and the surroundings.

Alcohol: An organic compound with an —OH group attached to a hydrocarbon in place of a hydrogen.

Aldehyde: An organic molecule with the structure

$$R-\overset{\overset{\displaystyle O}{\parallel}}{C}-H$$

Aliphatic hydrocarbons: Hydrocarbons that lack the benzene-ring structure.

Alkaline battery: A dry cell in which zinc serves as the anode and MnO_2 serves as the cathode in an alkaline KOH electrolyte.

Alkaline earth metals: The elements of Group IIA.

Alkali metals: The elements of Group IA, except hydrogen.

Alkane: A saturated hydrocarbon of general formula C_nH_{2n+2}.

Alkene: A hydrocarbon with one carbon-carbon double bond and having the general formula C_nH_{2n}.

Alkyl group: A group of atoms, derived from an alkane by loss of a hydrogen, that replaces a hydrogen in another molecule.

Alkyne: A hydrocarbon with one carbon-carbon triple bond and having the general formula C_nH_{2n-2}.

Allotropism: The existence of an element in two or more different forms.

Alloy: A solid solution of two or more metals.

Alpha amino acid (α-amino acid): A molecule with the structure

$$R-\underset{\underset{\displaystyle NH_2}{\mid}}{CH}-\overset{\overset{\displaystyle O}{\parallel}}{C}-OH$$

Alpha helix (α-helix): The coil structure assumed by a polypeptide chain.

Alpha particle (α-particle): The nucleus of a helium atom, $_2^4$He. One of the types of radiation given off by radioactive substances.

Alum: A double salt with the general formula $M^+M^{3+}(SO_4)_2 \cdot 12H_2O$, for example, $KAl(SO_4)_2 \cdot 12H_2O$.

Amalgam: A solution of a metal in mercury.

Amide: An organic molecule with the structure

$$R-\overset{\overset{\displaystyle O}{\|}}{C}-NH_2$$

Amine: An organic molecule derived from ammonia by replacing hydrogen atoms in NH_3 with organic groups.

Amorphous solid: A noncrystalline solid. It lacks the long-range order found in crystals. A glass. A supercooled liquid.

Ampere (A): The SI unit for electric current; one coulomb per second.

Amphiprotic: Able to either accept or donate a proton.

Amphoteric: Able to function as either an acid or a base, or to react with either an acid or a base.

Amplitude: The intensity of a wave; the maximum height of a peak as measured from the average height of peak and trough.

AMU: See Atomic mass unit.

Angstrom (Å): 1 Å = 10^{-8} cm = 10^{-10} m; 1 Å = 0.1 nm = 100 pm.

Anion: A negative ion.

Anode: In a gas discharge tube, it is the positive electrode. In electrochemistry, it is the electrode at which oxidation takes place. The anode carries a positive charge in electrolytic cells and a negative charge in galvanic cells.

Antibonding molecular orbital: A MO that removes electron density from between nuclei and destabilizes a molecule.

Aqua regia: One part concentrated HNO_3 and three parts concentrated HCl by volume.

Aromatic hydrocarbons: Hydrocarbons that contain a benzene-
   ring structure.

Arrhenius concept of acids and bases: Acids produce H_3O^+ in
   water, bases produce OH^- in water.

Arrhenius equation: The equation that relates the rate constant
   of a reaction, k, to the activation energy, E_a.
   $$k = Ae^{-E_a/RT}$$

Association: The joining together of solute molecules in a solution.

Asymmetric carbon atom: A carbon atom bonded to four different
   groups and which is a center for chirality.

Atomic emission spectrum: The spectrum emitted by atoms that
   have been energized. It consists of a relatively small number of different wavelengths of light. See also, Line spectrum.

Atomic mass unit (amu): One-twelfth of the mass of an atom of
   carbon-12.

Atomic number: The number of protons in the nucleus of a given
   atom.

Atomic radius: The effective radius of an atom.

Atomic weight: The relative average mass of the atoms of an element compared to carbon-12 as the standard.

Atomization energy: The amount of energy needed to convert
   one mole of a compound into neutral atoms. It is the total
   energy needed to break all the bonds in one mole of a compound.

Autoionization reaction: The reaction of a substance with itself
   to produce ions. For example, $H_2O + H_2O \rightleftharpoons H_3O^+ + OH^-$

Avogadro's number: The number of things (atoms, molecules, or
   whatever) in one mole. 6.022×10^{23} things = 1 mol

Avogadro's principle: Equal volumes of gases at the same temperature and pressure contain equal numbers of molecules.

Azeotrope: A mixture that has either a higher boiling point or a
   lower boiling point than either of the two pure components.

Azimuthal quantum number: ℓ, which for a given shell can have values of 0, 1, 2,..., n-1.

Balance: A device used to measure mass by comparison of the weights of two objects, one of known mass and the other of unknown mass.

Balanced equation: A chemical equation having the same number of atoms of each kind and the same net electrical charge on both sides of the arrow.

Band of stability: The collection of stable nuclei with various numbers of protons and neutrons that fall in a narrow band on a plot of number of neutrons versus number of protons.

Band theory of solids: A theory that explains the electrical behavior of solids in terms of bands of energy levels that extend throughout the entire crystal.

Barometer: A tube in which a column of mercury is supported by the atmospheric pressure and that is used to measure the pressure exerted by the atmosphere.

Base: In aqueous solutions, a substance that gives OH^-. Under the Brønsted-Lowry definition, a proton acceptor. Under the Lewis definition, an electron-pair donor. Under the solvent-system definition, a substance that yields the anion produced in the autoionization of a solvent.

Base ionization constant, K_b: The equilibrium constant for the ionization of a weak base.

Base units: The primary reference standards for the SI system.

Basic anhydride: A metal oxide, which reacts with water to give a metal hydroxide.

Basic oxygen process: A relatively fast method used to convert pig iron into steel.

Bessemer converter: An outdated method for converting pig iron into steel, which involves blowing air through the molten pig iron to reduce its carbon content.

Beta particle (β-particle): An electron given off by nuclei of radioactive substances. Its symbol is $_{-1}^{0}e$.

Bidentate ligand: A ligand that has two atoms that can become simultaneously attached to the same metal ion.

Bilayer structure: The structure of cell membranes in which the nonpolar tails of two layers of phospholipids face each other while the polar heads face the aqueous environment on either side of the bilayer.

Bimolecular collision: A collision between two molecules.

Binary acid: A substance with the general formula H_nX which produces acidic aqueous solutions (e.g., HCl, H_2S).

Binary compound: A compound composed of two elements (e.g., HCl, Na_2S, $FeCl_3$).

Binding energy: The energy equivalent of the mass defect. This energy would have been released on formation of a nuclide from its protons and neutrons.

Blast furnace: An apparatus that is used to reduce iron ore to metallic iron.

Body-centered cubic (bcc) unit cell: A cubic unit cell having atoms, molecules, or ions at the corners, plus one in the center of the cell.

Boiling point: The temperature at which the vapor pressure of a liquid is equal to the atmospheric pressure.

Boiling point diagram: A graph on which is plotted the boiling points of mixtures of varying composition. Also plotted is a curve showing the compositions of the vapor given off when solutions of different compositions boil. The vapor composition can be obtained from the liquid composition by a tie line running horizontally between the two curves.

Bomb calorimeter: A constant-volume calorimeter. Heats of reaction measured with this apparatus correspond to ΔE for the reaction.

Bond angle: When one atom forms a bond to each of two other atoms, the angle between the two bonds is the bond angle.

Bond distance: See Bond length.

Bond energy: The amount of energy needed to separate two bonded atoms to give electrically neutral particles.

Bond length: Also called bond distance. The distance between the nuclei of two atoms joined by a chemical bond.

Bond order: The <u>net</u> number of <u>pairs</u> of bonding electrons;
(no. of bonding e^- - no. antibonding e^-)/2.

Bonding molecular orbital: A MO that gives a buildup of electron
density between nuclei and helps stabilize a molecule.

Born-Haber cycle: A method of examining the contributing energy factors in an overall energy change. It involves constructing an alternative path from reactants to products and
analyzing the individual energy changes accompanying each
step along this alternative path.

Boyle's law: At constant temperature, for a fixed quantity of
gas, PV = constant.

Bragg equation: $n\lambda = 2d \sin \theta$. The equation is used to analyze
X-ray diffraction data obtained from crystals.

Branching step: A step in a chain reaction that produces more
free radicals than it consumes.

Breeder reactor: A nuclear reactor that produces plutonium from
^{238}U. The amount of fissionable Pu produced is greater
than the amount of nuclear fuel used to make it.

Brine: A concentrated aqueous sodium chloride solution.

Brønsted acid: Proton (H^+) donor.

Brønsted base: Proton (H^+) acceptor.

Buffer: A mixture that contains both a weak acid and a weak
base. It is capable of absorbing small additions of either
strong acid or strong base with little change in pH.

Buret: A graduated glass tube fitted with a valve (stopcock) at
one end that is used to dispense measured volumes of solutions.

Calcining: Heating a substance strongly in air.

Calorie (cal): The calorie is the amount of heat needed to raise
the temperature of 1 g of water by 1°C; 1 cal = 4.184 J,
exactly.

Calorimeter: A device used for the measurement of heats of
reaction.

Carbinol group: The group of atoms, C—OH.

Carbohydrate: A compound with the general formula, $C_n(H_2O)_m$.

Carbonyl group: The group, $\diagup C{=}O$

Carborundum: Silicon carbide, SiC, a common abrasive.

Carboxyl group: The group
$$\underset{\text{—C—OH}}{\overset{\text{O}}{\overset{\|}{}}}$$

Cast iron: Pig iron that has been cast into shapes by pouring the liquid metal into molds.

Catalyst: A substance that alters the rate of a reaction by providing a lower energy path (mechanism) from reactants to products. The catalyst is not used up as the reaction progresses.

Catenation: Linking together of atoms of the same element to give chains of atoms.

Cathode: In a gas discharge tube, it is the negative electrode. In electrochemistry, it is the electrode at which reduction takes place. The cathode carries a negative charge in electrolytic cells and a positive charge in galvanic cells.

Cathode rays: The stream of electrons that pass from cathode to anode in a gas discharge tube.

Cathodic protection: Protecting a metal from corrosion by coating it with another metal that is more easily oxidized and which thereby causes the protected metal to be the cathode in a galvanic cell.

Cation: A positive ion.

Cell potential, $\mathscr{E}_{cell}$: The emf that can be produced by any particular galvanic cell when no current is drawn from the cell.

Cell reaction: The overall chemical change that takes place in an electrolytic cell or a galvanic cell.

Celsius scale: A temperature scale on which water boils at 100°C and freezes at 0°C.

Change of state: Transformation of matter from one state to another, for example, from liquid to solid.

Charge cloud: Electron cloud.

Charge-to-mass ratio: The ratio of a particle's charge to its mass, expressed in units of coulombs per gram.

Charge transfer absorption band: A band of wavelengths absorbed by a substance from the electromagnetic spectrum. The photons that are absorbed cause a charge transfer process to occur.

Charge transfer process: The transfer of an electron from one atom to another brought about by absorption of a photon that has the energy required for the process.

Charles' law: At constant pressure, for a fixed quantity of gas, V/T = constant, where T is the temperature in kelvins.

Chelate: A complex ion containing rings formed by polydentate ligands.

Chemical bond: Forces of attraction that hold atoms together in compounds.

Chemical equation: A representation using chemical formulas of the changes that occur during a chemical reaction; a sort of "before-and-after picture" of the chemical system.

Chemical equivalence: A relationship between the amounts of two chemicals in a compound or a reaction. For example, in H_2O, 2.0 g H ~ 16 g O.

Chemical formula: A shorthand way of representing the composition of a substance using chemical symbols.

Chemical property: A statement of how a substance reacts chemically with some other substance.

Chemical symbol: The symbol that represents the name of an element. In formulas and equations it is used to represent an atom of the element.

Chemistry: The study of the composition of substances, the way their properties depend on their composition, and the way they interact with one another to form new materials.

Chiral: Possessing a "handedness." A term applied to optical isomers, which are nonsuperimposable on their mirror images.

Cholesteric liquid crystals: Rodlike molecules, similar in structure to cholesterol, that are arranged in layers in which the parallel rods in one layer are oriented in a diffcrent direction than the parallel rods in an adjoining layer.

Chromatography: A method of separating mixtures that relies on the different tendencies of substances to be adsorbed onto the surface of certain solids.

Cis-: A term applied to a geometrical isomer in which two groups are located on the same side of some central point in the molecule or ion.

Clathrate: Crystals in which noble gas atoms are trapped in a cagelike lattice.

Codon: A sequence of three bases along the m-RNA strand that serves as a code for a particular amino acid in a growing polypeptide chain.

Coefficients: Numbers that precede chemical formulas in a chemical equation.

Coenzyme: A substance needed by some enzymes in order to function.

Coinage metals: Copper, silver, and gold.

Coke: Coal that has had its volatile components driven off at high temperature. It is mostly carbon.

Colligative property: A property that depends only on the number of particles in a solution, and not on their chemical identity.

Collision theory: A theory of reaction rates that postulates that the rate of a reaction is proportional to the number of collisions that occur each second between the reactants.

Combined gas law: For a fixed quantity of gas, $\dfrac{P_1V_1}{T_1} = \dfrac{P_2V_2}{T_2}$

Common ion: An ion that is common to more than one salt. Na^+ is the common ion among $NaCl$ and $NaNO_3$.

Common ion effect: The solubility of a salt is less in a solution containing one of its ions than it is in pure water.

Complex ion (or simply a complex): A substance formed when one or more anions or neutral molecules become bonded to a metal ion.

Compound: A substance consisting of two or more elements combined in fixed proportions.

Compressibility: The ease with which a gas, liquid, or solid can be compressed to a smaller volume.

Concentrated: A large amount of solute dissolved in a solution.

Concentration: The amount of solute dissolved in a given amount of solvent or a given amount of solution.

Concentration cell: A cell in which both electrodes are composed of the same substance, but the ion concentrations in the two half-cells are different.

Condensation polymer: A polymer formed by the linking together of monomers, with the simultaneous elimination of small molecules, such as water.

Condensation reaction: The joining of two molecules, with the simultaneous elimination of a small molecule, such as water.

Conduction band: A band of energy levels in a crystal that is either empty or only partially filled, and which extends continuously through the crystal.

Conjugate acid: A substance related to another substance by the gain of a proton. (NH_4^+ is the conjugate acid of NH_3.)

Conjugate base: A substance related to another substance by the loss of a proton. (NH_3 is the conjugate base of NH_4^+.)

Constructive interference: The addition of intensities of waves that are in phase, so that the resultant wave is of greater amplitude.

Contact process: The commercial method of producing sulfuric acid from sulfur.

$$S + O_2 \longrightarrow SO_2$$
$$2SO_2 + O_2 \longrightarrow 2SO_3$$
$$SO_3 + H_2SO_4 \longrightarrow H_2S_2O_7$$
$$H_2S_2O_7 + H_2O \longrightarrow 2H_2SO_4$$

Continuous spectrum: An electromagnetic spectrum containing all wavelengths.

Conversion factor: A fraction formed from a valid relationship between two quantities, for example, $\dfrac{36 \text{ in.}}{1 \text{ yd}}$

Cooling curve: For a particular substance, a graph of temperature versus amount of heat removed. Temperatures corresponding to condensation of the vapor and freezing of the liquid can be read from the graph.

Coordinate covalent bond: A covalent bond in which both electrons are contributed by only one of the two joined atoms. Once formed, it is no different than any other covalent bond.

Coordination compound: A complex ion or its salt.

Coordination number: The number of donor atoms that surround a metal ion.

Copolymer: A polymer formed from two or more different kinds of monomer units.

Core electrons: The electrons in shells below an atom's outer shell.

Corrosion: The oxidation of a metal, which gives products that lack desirable metallic properties.

Coulomb: The SI unit of electrical charge. It is the amount of charge that passes a given point in a wire when a current of 1 ampere flows for 1 second.

Coulometer: An electrolysis cell in which the amount of chemical change that takes place is used to compute the number of coulombs that have passed through the cell.

Covalent bond: A chemical bond formed by the sharing of electrons between atoms.

Covalent crystal: A crystal in which lattice positions are occupied by atoms covalently bonded to other atoms at neighboring lattice sites.

Critical mass: The minimum amount of fissionable material that must be present to sustain a nuclear chain reaction.

Critical pressure, P_c: The vapor pressure of a substance at its critical temperature.

Critical temperature, T_C: The temperature above which a substance cannot exist as a separate liquid phase, regardless of the pressure.

Cross linking: The joining of adjacent polymer strands to give a three-dimensional rigid structure.

Cryolite: Na_3AlF_6, the solvent used in the Hall process.

Crystal field splitting, Δ: The difference in energy between sets of d orbitals in a complex.

Crystal field theory: A theory that considers the effects of the polar or ionic ligands of a complex on the energies of the d orbitals of the central metal ion.

Crystal lattice: The repeating symmetrical pattern of atoms, molecules or ions that occurs in a crystal.

Crystalline solid: A solid in which the particles are arranged in an orderly, repeating pattern.

Cyclo-: A prefix that means that the carbon chain in an organic compound exists in the form of a ring.

Cyclotron: A type of particle accelerator.

d-block elements: Transition elements in the main part of the periodic table which have partially completed d subshells.

Dalton: 1 dalton = 1 atomic mass unit

Dalton's law of partial pressures: For a mixture of gases, A, B, C, etc., the total pressure P_T is given by

$$P_T = p_A + p_B + p_C + \ldots$$

where p_A, p_B, etc. are partial pressures.

Data: The information obtained in an experiment.

Dative bond: A coordinate covalent bond.

Daughter isotope: An isotope produced in a radioactive decay.

Decay series: See Radioactive series.

Delocalized molecular orbital: A MO that spreads over more than two nuclei.

$\Delta G°$: See Standard free energy change.

$\Delta H°$: See Standard heat of reaction.

ΔH_{cryst}: See Molar heat of crystallization.

ΔH_{fus}: See Molar heat of fusion.

ΔH_{soln}: See Heat of solution.

$\Delta H_{sublimation}$: See Molar heat of sublimation.

$\Delta H_{vaporization}$: See Molar heat of vaporization.

Density: The ratio of an object's mass to its volume.

Deoxyribose: A five-carbon sugar that is one of the building blocks of DNA.

Desiccant: A drying agent.

Destructive interference: The cancellation of intensity that occurs when waves are out of phase, so that the resultant wave has diminished (or zero) amplitude.

Deuterium: An isotope of hydrogen; $_1^2H$, sometimes represented by the symbol, D.

Dextro-rotatory: An optical isomer that causes a rotation of plane polarized light in a clockwise direction, when looking toward the source, as the light passes through a solution of the isomer.

Dialysis: The passage of water and small molecules and ions, but not large molecules, through a membrane.

Diamagnetism: The nonattraction to a magnetic field experienced by substances having no unpaired electrons.

Differentiating solvent: A solvent that permits the observation of differences in the degree of acidity or basicity of solutes.

Diffraction: The scattering of light as it passes through a tiny pinhole or through a very narrow slit.

Diffraction pattern: The pattern that is produced by constructive and destructive interference of diffracted waves.

Diffusion: The mixing of two fluids, one into the other.

Difunctional molecule: A molecule having two functional groups.

Dilute: Very little solute dissolved in a solution.

Dimer: A particle formed by the joining together of two smaller particles.

Dipole: A molecule having partial positive and negative charges on opposite ends.

Dipole-dipole attractions: Attractions between molecules that are dipoles.

Dipole moment: The product of the partial charge on either end of a dipole multiplied by the distance between the partial charges. It is a measure of the extent of polarity of a bond.

Diprotic acid: An acid that can furnish two H^+ per molecule.

Disproportionation: A redox reaction in which a portion of a substance is oxidized while the rest is reduced. The same species undergoes both oxidation and reduction.

Dissociation: In general, the breaking apart of a substance into simpler substances. For solutions, the breaking apart of an ionic compound in water to give ions in solution. The term is also applied to the ionization of molecular compounds in water, which gives ions in the solution.

Dissociation constant: See Ionization constant.

Distillation: A means of separating mixtures that involves boiling the mixture and condensing the vapors of the more volatile component.

Disulfide bridge: A bridge between portions of polypeptide chains formed by $-S-S-$ which holds a polypeptide in its particular shape.

DNA: Deoxyribonucleic acid, the carrier of genetic information in the cell nucleus.

Domain: A region in a ferromagnetic substance in which the individual paramagnetic atoms are all aligned in the same direction.

Double bond: A covalent bond in which two pairs of electrons are shared.

Double helix: The intertwining of complementary DNA strands.

Double replacement reaction: A reaction between two salts in which cations and anions exchange partners, for example, $AgNO_3 + NaCl \longrightarrow AgCl + NaNO_3$.

Double salt: Crystals that contain the components of two different salts in a definite ratio.

Dry ice: Solid CO_2.

Ductility: A metal's ability to be drawn (stretched) into wire.

Dynamic equilibrium: An equilibrium in which two opposing processes are occurring at equal rates.

Effective nuclear charge: The effective charge experienced by a particular electron, which is a composite of the positive charge on the nucleus and the offsetting shielding of the nuclear charge by electrons in inner shells (and to some extent, by other electrons in the same shell).

Effusion: The escape of a gas under pressure through a very small opening into a region of low pressure.

Electric furnace: A furnace in which heat is generated by the passage of a heavy electric current through the contents of the furnace.

Electrochemical change: A chemical change that is caused by or that produces electricity.

Electrochemistry: The study of electrochemical changes.

Electrode: An electrically conducting substance that carries an electrical charge, either given to it by an external voltage source or acquired as the result of a chemical reaction (such as occurs in a battery).

Electrolysis: A chemical change caused by the passage of electricity through a molten ionic compound or through a solution that contains ions.

Electrolysis cell: An electrolysis apparatus.

Electrolyte: A substance that gives ions in aqueous solutions and thereby gives solutions that conduct electricity.

Electrolytic cell: An electrolysis apparatus.

Electrolytic conduction: The transport of charge through a solution by the movement of ions.

Electromagnetic radiation: General term used to describe light waves in all its various forms - e.g., X rays, ultraviolet and infrared radiation, visible light, TV waves, microwaves.

Electromotive force (emf): The voltage produced by a galvanic cell.

Electron: A subatomic particle that occurs outside the nucleus. It carries one unit of negative charge (-1.60×10^{-19} C) and has a mass (9.1×10^{-28} g) that is about 1/1800 of the mass of a proton. Its symbol is e^-.

Electron affinity (EA): The energy change that occurs when an electron is added to an isolated gaseous atom or ion (usually expressed in kJ/mol).

Electron capture: Capture of an electron from an atom's 1s orbital (K shell) by an unstable nucleus. It converts a proton to a neutron in the nucleus.

Electron cloud: Because of its wave properties, the electron is spread out like a cloud around the nucleus.

Electron configuration: The distribution of electrons in an atom's orbitals.

Electron density: The concentration of the electron's charge within a given volume.

Electron-dot formula: See Lewis structure.

Electron-pair bond: A covalent bond.

Electron spin: A property that the electron appears to have because it behaves like a tiny magnet.

Electronic structure: The distribution of electrons in an atom's orbitals.

Electronegativity: The relative attraction that an atom has for the electrons in a bond.

Electrophile: A Lewis acid, which seeks molecules with electron pairs that it can accept in the formation of a coordinate covalent bond.

Electrophilic displacement: An acid-base reaction in which one Lewis acid displaces another from its attachment to a Lewis base.

Electroplating: Depositing a thin metallic coating on an object by electrolysis.

Electropositive: Having a low electronegativity.

Electrovalent bond: An ionic bond.

Element: The simplest forms of matter that can exist under conditions normally encountered in a chemistry laboratory. Elements cannot be decomposed to simpler substances by chemical reactions.

Elementary process: One of the individual steps in a reaction mechanism.

Emf: See Electromotive force.

Empirical formula: Simplest whole-number ratio of atoms in a compound.

Enantiomers: Optical isomers.

End point: That point in a titration when the indicator changes color and the delivery of the titrant is halted.

Endothermic change: A change that absorbs energy from its surroundings.

Energy: The capacity to do work.

Energy level: A particular energy that an electron can have in an atom or molecule.

Enthalpy, H: Also called heat content. Defined by the equation, $H = E + PV$. At constant T and P, $\Delta H = \Delta E + P\Delta V$. ΔH is also the heat of reaction at constant pressure.

Enthalpy diagram: A diagram that displays in graphical form the enthalpy changes that accompany the various thermochemical equations that are combined to obtain some net change.

Enthalpy of formation: See Heat of formation.

Entropy, S: The thermodynamic quantity that describes the degree of randomness of a system. The greater the disorder or randomness, the higher is the statistical probability of the state and the higher is the entropy.

Enzyme: A biological catalyst that is very effective and highly specific for a particular reaction.

Enzyme-substrate electrode: A glass electrode coated with an enzyme-containing gel that makes the electrode sensitive to the concentration of the substance acted on by the enzyme.

Equation of state: An equation relating state variables.

Equation of state for an ideal gas: Ideal gas law, $PV = nRT$.

Equilibrium constant: The value that the mass action expression has when a chemical system is at equilibrium. It is K_p when partial pressures are used in the mass action expression, and it is K_c when molar concentrations are used.

Equilibrium law: An equation that sets the mass action expression equal to the equilibrium constant. A condition that must be fulfilled for a reaction to be at equilibrium.

Equilibrium vapor pressure: The pressure exerted by a vapor in equilibrium with its liquid.

Equilibrium vapor pressure of a solid: The pressure exerted by a vapor that is in dynamic equilibrium with a solid.

Equivalence point: That point in a titration when equal numbers of equivalents of reactants have been combined. Ideally, the equivalence point should occur at the end point during a titration.

Equivalent (eq): For acids or bases, the amount of substance that produces or reacts with one mole of H^+; for redox reactions, the amount of substance that gains or loses one mole of electrons.

Equivalent weight: The mass of one equivalent.

Ester: An organic molecule with the structure

$$R-\overset{\overset{\textstyle O}{\|}}{C}-O-R$$

Esterification: Reaction of an organic acid and an alcohol to form an ester.

Ether: An organic molecule with the structure $R-O-R$.

Evaporation: The conversion of a liquid to a gas.

Exact numbers: Numbers that come from definitions or a direct count of objects and contain an infinite number of significant figures because they contain no uncertainty.

Excluded volume: The volume that one molecule in a gas prevents other molecules from occupying.

Exothermic change: A change that releases energy to the surroundings.

Exponential notation: See Scientific notation.

Extensive properties: Properties that depend on the size of a sample.

Face-centered cubic (fcc) unit cell: A cubic unit cell having atoms, molecules, or ions at the corners and in the center of each face.

Factor-label method: The use of the cancellation of units associated with numbers in the solving of numerical problems.

Fahrenheit scale: A temperature scale defined by the boiling point of water = 212°F and the freezing point of water = 32°F.

Family of elements: A group in the periodic table.

Faraday ($\mathscr{F}$): One mole of electrons; 96,500 coulombs.

Fatty acid: An organic acid having a long hydrocarbon-chain tail.

Ferromagnetism: The strong magnetism associated with iron, cobalt, and nickel, which results from the alignment (in the solid state) of large numbers of paramagnetic atoms.

Filtrate: The liquid that passes through a filter.

First coordination sphere: The ligands that are attached to a metal ion in a complex.

First law of thermodynamics: A formal statement of the law of conservation of energy - the basis for Hess's law. $\Delta E = q - w$, where ΔE is the change in internal energy, q is the heat added to the system, and w is the work done by the system.

Fission: Splitting of a heavy unstable nucleus into several pieces, usually with the emission of large amounts of energy.

Flame test: Identification of an ion by the color produced when the ion is introduced into a flame. Sodium gives a yellow color.

Flotation: A method for concentrating sulfide ores of copper and lead.   Air is bubbled through a slurry of oil-coated ore particles, which stick to rising air bubbles and collect in the foam at the surface.

Formation constant: The equilibrium constant for the formation of a complex ion.

Formula unit: The collection of atoms specified by the chemical formula.   For example, a formula unit of $CaCl_2$ contains one calcium atom and two chlorine atoms.

Formula weight: The sum of the atomic weights of all the atoms in a formula unit.

Fractional crystallization: A procedure for purifying substances in which the impure solid is dissolved in a minimum amount of hot solvent.   The solution is then gradually cooled and crystals of the pure substance precipitate and are collected by filtration, while the impurities remain in solution.

Fractional distillation: A method used to separate mixtures of volatile liquids into their components.   It involves boiling the liquid, condensing the vapor, then boiling this condensed vapor, then condensing the new vapor, and so on.

Frasch process: A method of mining sulfur from deep wells. Compressed air and superheated water melt the sulfur and bring it to the surface.

Free radical: An extremely reactive chemical species that contains an unpaired electron.

Freezing point: The temperature at which an equilibrium can exist between liquid and solid at a particular pressure.

Frequency: The number of peaks in a wave that pass a given point per second.   The SI unit of frequency is the hertz; $1 \text{ Hz} = 1 \text{ s}^{-1}$.

Fuel cell: A galvanic cell in which the anode and cathode reactants can be fed continuously, so power can be drawn continuously.

Functional group: A group of atoms in an organic molecule that gives that molecule certain characteristic chemical properties.

Fundamental particle: A basic building block of all matter.   Examples are electrons, protons, and neutrons.

Fusion: Melting. In nuclear reactions, it is the joining of light-weight nuclei to produce heavier nuclei with the simultaneous emission of large amounts of energy.

Galvanic cell: An electrochemical cell in which a spontaneous redox reaction produces electricity.

Galvanizing: Coating a steel object with zinc.

Gamma rays (γ-rays): High energy, short wavelength radiation similar to X rays that is given off by radioactive substances.

Gangue: The unwanted rock and sand that is separated from an ore.

Gas discharge tube: A glass tube fitted with metal electrodes at either end and containing a gas at a low pressure. When a high voltage is placed across the electrodes, electric current passes through the tube and the gas glows.

Gay-Lussac's law: At constant volume for a fixed quantity of gas, P/T = constant, where T is the temperature in kelvins.

Gay-Lussac's law of combining volumes: When measured at the same temperature and pressure, the volumes of gases consumed or produced in a chemical reaction are in ratios of small whole numbers.

Geiger-Müller counter: A device used to detect radioactive emissions.

Genetic code: The base sequences found in m-RNA that specify the various amino acids used in building protein molecules.

Genetic disease: A "disease" or malfunction of cells that is caused by a fault in the DNA.

Geometrical isomers: Isomers that differ in the relative orientations of the atoms.

Gibbs free energy, G: A thermodynamic quantity that relates energy (enthalpy, H) and entropy, S. $G = H - TS$

Glass electrode: An ion-selective electrode consisting, generally, of a silver-silver chloride electrode immersed in an HCl solution which is separated from the bulk of the solution being tested by a thin glass membrane.

Glycoside linkage: The $-C-O-C-$ linkage that holds sugar molecules together in a polysaccharide.

Graham's law: The rate of effusion of a gas is inversely proportional to the molecular weight of the gas. Comparing two
gases, Rate (1)/Rate (2) = $\sqrt{M_2/M_1}$, where M is molecular
weight.

Gram: 1/1000 of the SI base unit for mass, the kilogram; about
1/30 of an ounce.

Ground state: Lowest energy state of an atom or molecule.

Group: A vertical column of elements in the periodic table.

Haber process: The process used in the commercial production
of ammonia: $N_2 + 3H_2 \longrightarrow 2NH_3$

Half-cell: A compartment of a galvanic cell in which either oxidation or reduction takes place.

Half-life, $t_{\frac{1}{2}}$: The time required for a reactant concentration to
be reduced to half of its initial value.

Half-reaction: An individual oxidation or reduction reaction that
includes the correct formulas for all species taking part in
the reaction, for example,
$2H_2O(\ell) \longrightarrow O_2(g) + 4H^+(aq) + 4e^-$.

Hall process: The method of producing aluminum by electrolysis
of Al_2O_3 dissolved in cryolite, Na_3AlF_6.

Halogens: The elements of Group VIIA.

Hard water: Water containing the ions Ca^{2+}, Mg^{2+}, and Fe^{3+}
which interfere with the action of soap by forming a precipitate with the soap.

Hardness ions: The ions found in hard water; Ca^{2+}, Mg^{2+}, and
Fe^{3+}.

Heat capacity: The amount of heat required to raise the temperature of a system by 1°C.

Heat content: Enthalpy.

Heat of solution: The amount of heat absorbed or released when
a given amount of solute dissolves to form a solution.

Heating curve: For a particular substance, a graph of temperature versus amount of heat added. Temperature correspond-
to melting and to boiling of the liquid can be read from the
graph.

Heavy water: Deuterium oxide, D_2O.

Heme group: A porphyrin structure having an Fe^{2+} ion in the center. It is responsible for hemoglobin's ability to carry oxygen and myoglobin's ability to hold oxygen until needed for metabolism.

Henry's law: $C_g = k_g p_g$, where C_g is the concentration of a gas dissolved in a solvent, p_g is the partial pressure of the gas over the solution, and k_g is a constant, called the Henry's law constant.

Hertz: The SI unit of frequency; $1 \text{ Hz} = 1 \text{ s}^{-1}$.

Hess's law of heat summation: When we add thermochemical equations to obtain some net change, we add their corresponding heats of reaction to obtain the heat of reaction for the net change. Also,
$$\Delta H_{reaction} = (\text{sum } \Delta H_f^\circ \text{ products}) - (\text{sum } \Delta H_f^\circ \text{ reactants})$$

Heterocycle: An organic ring structure in which one or more atoms in the ring is an element other than carbon.

Heterogeneous catalyst: A catalyst that is in a different phase than the reactants. The reactants are adsorbed on the catalytic surface where the reaction occurs.

Heterogeneous equilibrium: An equilibrium in which the reactants and products are not all present in the same phase.

Heterogeneous mixture: A mixture in which different portions have different properties.

Heterogeneous reaction: A reaction in which the reactants and/or products are in different phases.

High spin complex: A complex in which there is a minimum pairing of electrons in the central metal ion.

Homogeneous catalyst: A catalyst that is in the same phase as the reactants.

Homogeneous mixture: A mixture having the same uniform properties throughout.

Homogeneous reaction: A reaction in which all of the reactants and products are in the same phase.

Homologous series: A series of hydrocarbons in which each member differs from the preceding member by the same grouping of atoms.

Hund's rule: The lowest energy electron configuration results when electrons that occupy orbitals of equal energy are spread out over the orbitals as much as possible with spins unpaired.

Hybrid atomic orbitals: Orbitals formed by mixing the basic atomic orbitals of an atom. They are more effective at overlap than ordinary atomic orbitals.

Hydrate: A crystal that contains molecules of water in fixed proportions relative to the other substances present.

Hydrated ion: An ion that has become surrounded by water dipoles, to which it is attracted.

Hydration: The surrounding of an ion or molecule by molecules of water.

Hydration energy: The amount of energy liberated when an ion (or other solute particle) becomes surrounded by water molecules. It is usually expressed in units of kJ/mol of solute hydrated.

Hydride ion: H^-

Hydrides: Binary compounds containing hydrogen.

Hydrocarbon: A compound composed of only carbon and hydrogen.

Hydrogen bond: An extra strong dipole-dipole attraction that occurs between molecules in which hydrogen is covalently bonded to nitrogen, oxygen, or fluorine.

Hydrogen economy: An economy built around the extensive use of hydrogen as a fuel.

Hydrogen electrode (standard hydrogen electrode): The standard of comparison for reduction potentials,
$2H^+(aq) + 2e^- \rightleftharpoons H_2(g)$, $\mathcal{E}^\circ_{H^+} = 0.00$ V at 25°C, 1 atm, and 1 M H^+.

Hydrogenation: Addition of H_2 to organic molecules having double bonds.

Hydrolysis: Reaction of a substance with water. Hydrolysis of anions produces basic solutions; hydrolysis of cations produces acidic solutions.

Hydrolysis constant (K_h): The equilibrium constant for the hydrolysis of an anion or cation.

$$K_h = \frac{K_w}{K_a} \text{ or } \frac{K_w}{K_b}$$

Hydronium ion: H_3O^+

Hypothesis: A tentative explanation of the results of one or more experiments.

Ideal gas: A hypothetical gas that would obey the gas laws perfectly under all conditions.

Ideal gas law: $PV = nRT$

Ideal solution: A solution for which $\Delta H_{soln} = 0$, and in which solute-solute, solvent-solvent, and solute-solvent attractions are all the same.

In phase: A condition that is met when the peaks (and troughs) of waves coincide.

Indicator: A substance that changes color to signal the completion of a reaction during a titration. For acid-base titrations, a weak acid or weak base whose molecular form differs in color from its ionic form.

Induced dipole: A dipole created when the electron cloud of an atom or a molecule is distorted by a neighboring dipole or by an ion.

Inhibition: The blocking of enzyme activity by blockage of the active enzyme site.

Inhibition step: In a chain reaction, a step in the mechanism that removes product molecules and thereby "inhibits" the overall production of the products.

Inhibitor: A substance that blocks the action of a catalyst.

Initiation step: The first step in a chain reaction in which a reactant molecule is converted to one or more free radicals.

Inner orbital complex: A complex in which the metal ion uses d orbitals below its outer shell in forming hybrid orbitals used for bonding to the ligands.

Inner transition elements: The two long rows of elements below the main body of the periodic table.

Inorganic compound: Compounds whose structures are not primarily determined by the linking together of carbon atoms.

Instability constant: The equilibrium constant for the decomposition of a complex ion into its components.

Instantaneous dipole: A momentary dipole caused by the erratic movement of electrons.

Intensive properties: Properties that are independent of the size of a sample.

Intermolecular forces of attraction: Attractive forces that occur between molecules.

Internal energy, E: The total kinetic and potential energy possessed by a system.

International System of Units: A modified metric system adopted for worldwide use in 1960 by the General Conference of Weights and Measures.

Interstitial solid solution: A solid solution in which the solute atoms fit into empty spaces between atoms of the solvent in the solvent's lattice.

Ion: An electrically charged atom or group of atoms.

Ion-electron method: A method for balancing redox reactions that uses half-reactions.

Ion product: The product of ion concentrations. For a salt, the product of the concentrations of the ions, each raised to an exponent equal to the number of ions of that kind produced by one formula unit of the salt (e.g., for Ag_2S, the ion product is $[Ag^+]^2[S^{2-}]$).

Ion product constant: An equilibrium constant that is equal to a product of ion concentrations; for example, $K_W = [H^+][OH^-]$.

Ion-selective electrode: An electrode that is sensitive to the concentration of one particular kind of ion.

Ionic bond: The attractions between ions that hold them together in ionic compounds.

Ionic compound: A compound composed of positive and negative ions (e.g., NaCl).

Ionic crystal: A crystal that has ions located at the lattice points.

Ionic equation: A chemical equation in which all water-soluble electrolytes are written in ionic form, while solids and weak electrolytes are written in molecular form.

Ionic potential: The ratio of an ion's charge, q, to its radius, r; $\phi = q/r$

Ionic radius: The effective radius of an ion, which is nearly constant from compound to compound.

Ionization constant: The equilibrium constant for the ionization of a weak electrolyte.

Ionization energy (IE): The energy needed to remove an electron from an isolated gaseous atom, ion, or molecule (usually expressed in units of kJ/mol).

Ionization isomers: Coordination compounds that have the same formula but in which different anions serve as ligands.

Isomers: Compounds having the same formula but differing in the way their atoms are arranged.

Isothermal: A change that occurs at constant temperature.

Isotonic solutions: Solutions having the same osmotic pressure.

Isotope dilution: An analytical method that uses the extent to which a radioisotope undergoes dilution when added to a sample to compute the original amount of nonradioactive isotope in the sample.

Isotopes: Atoms of the same element that differ slightly in their masses. The nuclei of all the isotopes of a given element have the same number of protons, but they have different numbers of neutrons.

Joule (J): The SI unit of energy; $1 \text{ J} = 1 \text{ kg m}^2/\text{s}^2$; the kinetic energy possessed by an object with a mass of 2 kg moving at a speed of 1 m/s.

K capture: See Electron capture.

Kelvin (K): The SI unit of temperature, equal in size to one
    Celsius degree.

Kelvin temperature: $K = °C + 273.15$

Ketone: An organic molecule with the structure

$$R-\overset{\overset{\displaystyle O}{\|}}{C}-R$$

Kilocalorie (kcal): 1 kcal = 1000 cal

Kilojoule (kJ): 1 kJ = 1000 J

Kinetic energy: Energy that an object possesses because of its
    motion; $KE = \frac{1}{2}mv^2$.

Kinetic molecular theory: The model of a gas that explains the
    properties of a gas and accounts for the gas laws. It states
    that an ideal gas is composed of point-sized particles in
    rapid random motion and that the temperature of the gas is
    directly proportional to the average kinetic energy of its
    particles.

Lanthanide contraction: The decrease in size experienced by
    elements 58 through 71 which causes the elements that follow
    the lanthanides to have unusually small sizes and large ion-
    ization energies.

Lanthanide elements: Elements 58 - 71.

Lattice: A repeating pattern of points.

Lattice energy: Energy released by the imaginary process in
    which isolated ions come together to form a crystal of an
    ionic compound.

Law: A statement of behavior based on the results of many ex-
    periments. Laws are often expressed in equation form.

Law of conservation of energy: Energy is neither created nor
    destroyed but, instead, can only be transformed from one
    kind of energy to another.

Law of conservation of mass: Mass is neither created nor de-
    stroyed in a chemical reaction.

Law of definite composition: See Law of definite proportions.

Law of definite proportions: In a pure chemical substance, the elements are always present in definite proportions by mass.

Law of mass action: At equilibrium at a particular temperature, the mass action expression for a particular reaction will always equal the same value. There are no restrictions on the individual concentrations other than that they satisfy this relationship.

Law of multiple proportions: When two compounds are formed from the same two elements, the masses of one element in the two compounds that combine with the same mass of the other element are in a ratio of small whole numbers.

Le Châtelier's principle: When a system that is in dynamic equilibrium is subjected to a disturbance that upsets the equilibrium, the system undergoes a change that counteracts the disturbance and restores equilibrium.

Lead storage battery: The common automobile battery, in which during discharge Pb serves as the anode and PbO_2 serves as the cathode in an H_2SO_4 electrolyte.

Leveling effect: Differences in acid strength are "leveled out" by dissolving acids in basic solvents. A similar effect occurs when bases are dissolved in acidic solvents.

Leveling solvent: A solvent that, because of its acidity or basicity, obscures the differences in the strengths of bases or acids, respectively.

Levo-rotatory: An optical isomer that causes a rotation of plane polarized light in a counterclockwise direction, when looking toward the source, as the light passes through a solution of the isomer.

Lewis acid: An electron-pair acceptor.

Lewis base: An electron-pair donor.

Lewis structure: Also called Lewis formula or electron-dot formula. A structural formula drawn with Lewis symbols that shows the valence electrons using dots and dashes.

Lewis symbol: The symbol of an element surrounded by dots that represent the valence electrons of an atom of that element.

Ligand: A molecule or anion that can bind to a metal ion to form a complex.

Limiting reactant: The reactant that is the first to be completely consumed in a chemical reaction; the reactant that limits the amount of product formed in a reaction.

Line spectrum: The spectrum, consisting of only a relatively few wavelengths, that is produced when the light emitted by energized or excited atoms is passed first through a thin slit, then through a prism, and then allowed to fall on a screen or a piece of photographic film. Also called an atomic emission spectrum.

Linear molecule: A molecule in which all the atoms lie along a straight line.

Lipid: A water-insoluble substance that can be extracted from cells by nonpolar solvents.

Liquid crystal: A substance that is able to flow like a liquid but which has some physical properties normally associated with crystals.

Liter: A unit of volume equal to 1 dm^3; 1.057 quarts.

London forces: Weak attractive forces caused by instantaneous dipole-induced dipole attractions.

Lone pair: A pair of electrons in the valence shell of an atom that is not shared with another atom. An unshared pair of electrons.

Low spin complex: A complex in which there is maximum electron pairing in the orbitals of the central metal ion.

Magic numbers: Numbers of protons and neutrons found in especially stable nuclei.

Magnetic quantum number: m, which can have integer values from $-\ell$ to $+\ell$.

Main transition elements: See d-block elements.

Malleability: A metal's ability to be hammered or rolled into thin sheets.

Manganese nodules: Lumps the size of an orange that contain large concentrations of manganese and iron. They are found on the ocean floor.

Manometer: A U-shaped tube, filled partially with a liquid, that is used to measure pressures of trapped gases.

Markovnikov's rule: During an addition reaction, the hydrogen of the molecule being added becomes attached to the carbon that is already bonded to the most hydrogen atoms.

Mass: A measure of the amount of matter in an object; a measure of an object's resistance to a change in velocity.

Mass action expression: A fraction that can be constructed from the overall balanced equation for a reaction. For the general reaction, $aA + bB \rightleftharpoons mM + nN$, the mass action expression is

$$\frac{[M]^m [N]^n}{[A]^a [B]^b}$$

For reactions involving gases, partial pressures can be used in place of molar concentrations.

Mass defect: The difference between the actual mass of a nuclide and the mass computed by adding the weights of the corresponding numbers of protons and neutrons.

Mass number: The sum of the numbers of protons and neutrons in the nucleus of an atom.

Mass spectrometer: A device that allows the determination of the charge-to-mass ratio of positive particles.

Matter: Anything that occupies space and has mass.

Mean free path: The average distance traveled by a molecule between collisions.

Mechanism: The series of individual steps in a chemical reaction that gives the net, overall change.

Melting point: The same temperature as the freezing point of a substance.

Messenger RNA (m-RNA): A form of RNA that carries the code for the sequence of amino acids in proteins from the DNA inside the cell nucleus to locations outside the nucleus where protein synthesis takes place.

Meta-: In the benzene ring, positions separated by one intervening carbon atom.

Metal carbonyl compound: A compound formed from a metal and carbon monoxide, for example, $Ni(CO)_4$.

Metallic conduction: The transport of charge through a metal by the movement of electrons.

Metallic crystal: A solid having positive ions at the lattice positions which are attracted to a "sea of electrons" that extends throughout the entire crystal.

Metalloids: Elements with properties that lie between those of metals and nonmetals. They are located around the diagonal line running from boron (B) to astatine (At) in the periodic table.

Metallurgy: The science and technology of metals; it is concerned with the procedures and chemical reactions that are used to separate metals from their ores and make them ready for practical uses.

Metals: Elements that are lustrous, have high thermal and electrical conductivity, and that are generally ductile and malleable. In compounds they nearly always have positive oxidation states. In the periodic table, they are located to the left of the diagonal line running from boron to astatine.

Metathesis reaction: See Double replacement reaction.

Meter: The SI base unit for length; 39.37 inches.

MeV (million electron volts): An energy unit used to express binding energies and energy changes associated with nuclear reactions. 1 MeV per nuclide corresponds to an energy of 9.65×10^7 kJ/mol.

Micelle: The grouping together of fatty acid anions with their nonpolar "tails" intermingling and their anionic "heads" facing out toward the aqueous environment.

Miscible: Two liquids are miscible if they are soluble in each ohter in all proportions.

Mixture: Two or more substances that may be combined in variable proportions.

Molal boiling point elevation constant: A constant, characteristic of each particular solvent, that relates the boiling point elevation for a solution to the molal concentration of solute.

Molal freezing point depression constant: A constant, characteristic of each particular solvent, that relates the freezing point depression for a solution to the molal concentration of solute.

Molality: A concentration unit that is the ratio of the number of moles of solute to the number of kilograms of solvent.

Molar: A term that means "moles of solute/liter of solution."

Molar concentration: A ratio of moles of solute to liters of solution; the number of moles of solute per liter of solution.

Molar heat capacity: The amount of heat required to raise the temperature of one mole of a substance by 1°C.

Molar heat of crystallization, ΔH_{cryst}: The heat evolved when one mole of a liquid freezes. It is equal in magnitude, but opposite in sign to the molar heat of fusion.

Molar heat of fusion: The amount of heat absorbed when one mole of a solid melts to give one mole of liquid at constant temperature and pressure.

Molar heat of sublimation: The amount of heat absorbed when one mole of solid sublimes to give one mole of vapor at constant temperature and pressure.

Molar heat of vaporization: The amount of heat absorbed when one mole of liquid is converted to one mole of vapor at constant temperature and pressure.

Molar solubility: The number of moles of solute required to give 1 liter of a saturated solution of the solute.

Molarity: See Molar concentration.

Mole (mol): 6.022×10^{23} things. The formula weight of a substance in grams.

Mole fraction: A concentration unit that is the ratio of the number of moles of a given component to the total number of moles of solution.

Mole percent (mol %): mole percent = 100 x mol fraction

Molecular crystal: A crystal that has molecules or individual atoms at the lattice points.

Molecular equation: A chemical equation in which the reactants and products are written as if they were molecular.

Molecular formula: A chemical formula that specifies the number of atoms of each kind present in a molecule.

Molecular orbital theory: Theory of covalent bonding that views a molecule as a collection of positive nuclei surrounded by a set of bonding and antibonding molecular orbitals of different energies.

Molecular orbitals: Orbitals that extend over two or more atomic nuclei.

Molecular weight: The sum of the atomic weights of all the atoms in a molecule.

Molecule: An electrically neutral group of atoms bound tightly enough together that they behave as, and can be recognized as, a single particle.

Monodentate ligand: A ligand that can attach itself to a metal ion by only one atom.

Monomer: A small molecule that combines with others to form polymers.

Monoprotic acid: An acid that can furnish one H^+ per molecule.

Monosaccharide: The simple units that combine to form poly-saccharide chains in starch and cellulose.

n-type semiconductor: A semiconductor in which the charge is transported by electrons.

Negative deviations from Raoult's law: Solutions that exhibit negative deviations have vapor pressures that are lower than predicted by Raoult's law.

Nematic liquid crystal: Composed of long rodlike molecules packed like short pieces of uncooked spaghetti.

Nernst equation: $$\mathscr{E}_{cell} = \mathscr{E}^{\circ}_{cell} - \frac{0.0592}{n} \log Q$$

Net bond order: $\dfrac{\text{(no. of bonding e}^-) - \text{(no. of antibonding e}^-)}{2}$

Net ionic equation: A chemical equation obtained by omitting spectator ions from an ionic equation. It shows the net chemical change that occurs.

Neutralization: The reaction of an acid with a base. In aqueous solutions, the products are a salt plus water.

Neutron: A subatomic particle found in the nuclei of atoms. It is electrically neutral and has a mass (1.67×10^{-24} g) that is just slightly larger than that of a proton.

Neutron activation analysis: An analytical procedure in which nonradioactive isotopes are bombarded by neutrons, making some of their atoms radioactive. From the frequencies and intensities of the γ-rays emitted by the bombarded sample, the concentrations of the various elements in the sample can be calculated.

Nickel-cadmium cell: The nicad battery, in which during discharge Cd serves as the anode and NiO_2 serves as the cathode in an alkaline electrolyte.

Noble gases: Elements of Group 0 in the periodic table; helium, neon, argon, krypton, xenon, and radon.

Noble metal: A metal that is very unreactive, for example, gold and platinum.

Node: A place where the amplitude or intensity of a wave is zero.

Nonelectrolyte: A substance that does not dissociate into ions in water.

Nonmetals: Elements that are poor conductors of heat and electricity and lack the other properties of metals. In the periodic table, they are located to the right of the diagonal line running from boron to astatine.

Nonstoichiometric compound: A substance in which the elements are combined in a not quite whole number ratio by moles.

Normal boiling point: The temperature at which the vapor pressure of a liquid equals 1 atm.

Normality: The number of equivalents of solute per liter of solution.

Nuclear transformations: Changes in nuclei brought about by bombarding them with high energy particles.

Nucleic acid: A polymer of nucleotide units.

Nucleophile: A Lewis base, which seeks nuclei to which it can donate an electron pair in the formation of a coordinate covalent bond.

Nucleophilic displacement: An acid-base reaction in which one Lewis base displaces another from its attachment to a Lewis acid.

Nucleotide: The units that form the building blocks of nucleic acids. They are composed of even simpler units - a five-carbon sugar, phosphoric acid, and a nitrogenous base.

Nucleus: The very tiny, massive particle found at the center of an atom. It contains all of the atom's positive charge and nearly all of its mass. Protons and neutrons are found in the nucleus.

Nuclide: A general term used to describe the nucleus of a particular isotope.

Octahedral molecule: A molecule in which the central atom is bonded to six others that are located at the vertices of an octahedron (a figure consisting of two square pyramids sharing a common square base).

Octet rule: An atom tends to gain or lose electrons until its outer shell consists of eight electrons.

Olefin: An alkene.

Open hearth furnace: A furnace used to convert pig iron into steel.

Optical activity: The rotation of the plane of polarized light as it passes through a chiral substance.

Optical isomers: Isomers that are nonsuperimposable mirror images of each other.

Orbital: A particular electron waveform with a particular energy. In an atom, each orbital has a specific set of values for n, ℓ, and m.

Orbital diagram: A diagram showing an atom's orbitals in which the electrons are represented by arrows to indicate paired and unpaired spins.

Order (of reaction): The sum of the exponents in the rate law is the overall order of the reaction. Each exponent gives the order with respect to a certain reactant.

Ore: A substance that contains a desirable constituent in concentrations large enough to make its recovery economically worthwhile.

Organic acid: An organic molecule with the structure

$$R-\overset{\overset{\displaystyle O}{\|}}{C}-OH$$

Organic compounds: Compounds whose structures are determined primarily by the linking together of carbon atoms. They are hydrocarbons or can be considered to be derived from hydrocarbons.

Ortho: In the benzene ring, positions that are adjacent to each other.

Osmosis: The selective passage of solvent through a semipermeable membrane from a solution of low solute concentration to a solution of high solute concentration.

Osmotic pressure: The pressure that must be exerted on the more concentrated solution to prevent osmosis from occurring.

Ostwald process: The commercial process used to make nitric acid from ammonia.

$$4NH_3 + 5\,O_2 \longrightarrow 4NO + 6H_2O$$
$$2NO + O_2 \longrightarrow 2NO_2$$
$$3NO_2 + H_2O \longrightarrow 2HNO_3 + NO$$

Out of phase: A condition that is met when the peak of one wave coincides with the trough of another wave.

Outer orbital complex: A complex in which the metal ion uses d orbitals in its outer shell in forming hybrid orbitals used for bonding to the ligands.

Overlap of orbitals: A portion of two orbitals from different atoms share the same space.

Oxidation: Loss of electrons; increase in oxidation number.

Oxidation number: The charge an atom in a compound would have if all of the electrons in the bonds belonged entirely to the more electronegative atoms.

Oxidation number change method: A method for balancing redox reactions that makes the total increase in oxidation number equal to the total decrease in oxidation number.

Oxidation-reduction reaction: A reaction involving the transfer of electrons from one species to another.

Oxidation state: Means the same as oxidation number.

Oxidizing agent: Substance that causes oxidation; it is reduced.

Oxoacid: An acid containing hydrogen, oxygen, and another element.   (e.g., HNO_3, H_3PO_4, H_2SO_4)

Ozone: O_3

p-type semiconductor: A semiconductor in which the charge is transported by the movement of positive "holes," or electron vacancies.

Paired electrons: For each electron with $s = +\frac{1}{2}$, there is an electron with $s = -\frac{1}{2}$.

Pairing energy: The amount of energy that must be absorbed to cause two electrons to occupy the same orbital with their spins paired.

Para-:   In the benzene ring, positions separated by two intervening carbon atoms.

Paraffin: An alkane.

Paramagnetism: Weak magnetism (attraction toward a magnetic field) that a substance has when it contains unpaired electrons.

Parent isotope: The isotope that decays in a radioactive decay.

Partial charge: Charges at opposite ends of a dipole that are less than full 1+ or 1- charges.

Partial pressure: The pressure that is exerted by a particular gas in a mixture of gases.

Particle accelerator: Instruments that accelerate charged particles to very high speeds (high energies) before allowing them to bombard target nuclei.

Parts per billion (ppb): The number of grams of the component in question per billion grams of solution.

$$\text{ppb } X = \frac{\text{grams } X}{\text{grams solution}} \times 10^9$$

Parts per million (ppm): A concentration unit that specifies the number of grams of the component in question per 1 million grams of solution. It is used to express the concentrations of very dilute mixtures.

$$\text{ppm } X = \frac{\text{grams } X}{\text{grams solution}} \times 10^6$$

Pascal: The SI unit of pressure; a pressure of one newton per square meter; $1 \text{ Pa} = 1 \text{ N/m}^2$.

Pauli exclusion principle: No two electrons in the same atom can have the same values for all four of their quantum numbers.

Peptide: A polymer of α-amino acids.

Peptide bond: The structure,

$$\cdots-\overset{\overset{\textstyle O}{\|}}{C}-\underset{\underset{\textstyle H}{|}}{N}-\cdots$$

Peptide linkage: See Peptide bond.

Percent by weight (% w/w): The number of grams of the component in question per 100 grams of solution.

$$\% \ X \ (w/w) = \frac{\text{grams } X}{\text{grams solution}} \times 100$$

Percent dissociation:

$$\% \text{ dissociation} = \frac{\text{amount dissociated}}{\text{total amount available}} \times 100$$

Percent uncertainty: The estimated uncertainty in a measurement divided by the value of the measurement, and then this quotient multiplied by 100.

Percentage composition: The percents by weight of the elements in a compound.

Percentage yield: $\dfrac{\text{actual yield}}{\text{theoretical yield}} \times 100$

Period: A horizontal row of elements in the periodic table.

Periodic law: When the elements are arranged in order of increasing atomic number, there is a periodic recurrence of properties.

pH: A logarithmic measure of acidity ($pH = - \log [H^+]$). A solution is acidic if its pH < 7, neutral if its pH = 7, and basic if its pH > 7.

Phase: A homogeneous region within a sample.

Phase change: Transition of a sample from solid to liquid, liquid to gas, solid to gas, or vice versa.

Phase diagram: A pressure-temperature graph on which are plotted temperatures and pressures at which equilibrium exists between the states of a substance. It defines T-P regions in which the solid, liquid, and gaseous states of the substance can exist.

Phenyl group: Benzene, with one hydrogen removed, that becomes attached to some other organic molecule in place of a hydrogen.

Phlogiston: An imaginary substance once thought to be lost by an object when it burned.

Phospholipid: An ester of glycerol, two fatty acids, and phosphoric acid (which in turn is esterified to another alcohol). Phospholipids are found in cell membranes.

Photochemical smog: A type of urban air pollution caused by the interaction of sunlight and nitrogen oxides, which produces ozone and the unpleasant products of the reaction of ozone and airborn hydrocarbons.

Photon: A tiny packet of electromagnetic energy (light energy) whose energy is given by $E = h\nu$.

Physical process: A change that does not alter the chemical properties of the substances involved.

Physical property: A property (e.g., color) that can be specified without reference to any other substance.

Pi bond (π-bond): A bond formed by the sideways overlap of a pair of p orbitals. Electron density is concentrated in two separate regions that lie on opposite sides of the imaginary line joining the nuclei.

Pickling: The removal of rust from iron or steel by reaction with acid.

Pig iron: The impure iron that comes from the blast furnace.

pK_W: $-\log K_W = 14.00$ (at 25°C)

Planar triangular molecule: A molecule in which three atoms surround the central atom at the corners of a triangle.

Planck's constant: A constant, h, that permits the calculation of the energy of a photon of frequency ν by the equation, $E = h\nu$. $h = 6.6262 \times 10^{-34}$ J s.

Plasma: A very hot, high energy "gas" composed of ions.

pOH: $-\log [OH^-]$

Polar bond: See Polar covalent bond.

Polar covalent bond: A covalent bond in which more than half of the bond's negative charge is concentrated around one of the two atoms.

Polarizability: The ease of distortion of a molecule's electron cloud.

Polarized light: Light in which the vibrations are all in the same plane.

Polyatomic ion: An ion composed of two or more atoms.

Polydentate ligand: A ligand that has two or more atoms that can become simultaneously attached to a metal ion.

Polyester: A copolymer of a difunctional alcohol with a difunctional organic acid.

Polypeptide: A peptide composed of many α-amino acid units.

Polyprotic acid: An acid that can furnish more than one H^+ per molecule.

Polysaccharide: A polymer formed from simple, monomeric sugar molecules.

Porphyrin: A square planar ligand structure found in heme and chlorophyll.

Position of equilibrium: The relative proportions of reactants and products in an equilibrium system.

Positive deviations from Raoult's law: Solutions that exhibit positive deviations have vapor pressures higher than predicted by Raoult's law.

Post-transition element: In the periodic table, an element that occurs after a row of transition elements.

Post-transition metal: A metal that occurs in the periodic table immediately following a row of transition elements.

Potential energy: Energy that an object possesses because of the attractive and/or repulsive forces that it experiences.

Potentiometer: A device that permits the measurement of the emf of a galvanic cell without drawing current from the cell.

Precipitate: A solid that forms in a solution.

Precision: How closely measurements of the same quantity come to each other.

Primary alcohol: An alcohol with the formula $R-CH_2OH$.

Primary structure: The sequence of the various α-amino acids in a polypeptide.

Primitive lattice: A lattice in which the unit cell has lattice points only at the corners.

Principal quantum number: n, which can values of 1, 2, 3,...,∞

Probability distribution: The way that the probability of finding the electron varies from place to place around the nucleus of an atom.

Products: Substances that appear on the right side of a chemical equation; the substances that are present after a chemical reaction has occurred.

Propagation step: A step in a chain reaction in which a free radical reacts with a reactant molecule to give a product molecule and another free radical.

Properties: Characteristics of matter that serve to identify it.

Protein: A polymer of α-amino acids having a specific biological function.

Protium: $_1^1H$, the common isotope of hydrogen.

Proton: A subatomic particle found in the nuclei of all atoms. It carries one unit of positive charge ($+1.60 \times 10^{-19}$ C) and its mass (1.67×10^{-24} g) is about 1800 times larger than that of the electron. It is often symbolized as H^+ in chemical reactions.

Pseudonoble gas configuration: For a given n, the configuration $ns^2np^6nd^{10}$.

Quanta: Tiny packets of light energy having energy $E = h\nu$. Also called photons.

Qualitative observation: An observation made without measurement of numerical values.

Quantitative observation: An observation involving measurements that result in numerical data.

Quantum mechanics: See Wave mechanics.

Quantum number: A number related to the energy, shape, or orientation of an orbital. Also, a number related to the spin of the electron.

Quaternary structure: The way certain folded polypeptide chains pack together.

Racemic: A mixture that contains equal numbers of the two optical isomers of a substance, and which therefore does not rotate the plane of polarized light in either direction.

Radioactive decay: The gradual transformation of a collection of unstable nuclei into a collection of stable nuclei by the emission of various forms of radiation (α, β, or γ).

Radioactive series: A series of radioactive isotopes produced one from another, finally terminating in the formation of a stable, nonradioactive isotope.

Radioactivity: The spontaneous emission of radiation by certain unstable atomic nuclei.

Radioisotope: A radioactive isotope.

Radionuclide: Nucleus of a radioactive isotope.

Raoult's law: $P_A = X_A P_A^\circ$, where P_A° is the vapor pressure of pure A, X_A is the mole fraction of A in a solution, and P_A is the vapor pressure of A over the solution.

Rare earth metals: The lanthanides.

Rate: A ratio in which units of time appear in the denominator, for example, 40 miles/hr or 3.0 mol per liter/second.

Rate constant: The proportionality constant in the rate law; the rate of reaction when all reactant concentrations are 1 M.

Rate-determining step: The slow step in a reaction mechanism which determines how fast the products appear.

Rate law: An equation that relates the rate of a reaction to the molar concentrations of the reactants raised to powers.

Rate of reaction: How quickly the reactants disappear and the products form, expressed in units of mol liter^{-1} s^{-1}.

Reactants: Substances that appear on the left side of a chemical equation; the substances that are present before a chemical reaction occurs.

Reaction coordinate: The path of the reaction: molecules coming together in a collision, followed by bonds being broken as new ones are formed, and then the moving apart of the newly formed product molecules.

Reaction quotient: See Mass action expression.

Reaction rate: See Rate of reaction.

Red phosphorus: A relatively unreactive form of phosphorus, which has an unknown molecular structure.

Redox: A term meaning oxidation-reduction.

Reducing agent: Substance that causes reduction; it is oxidized.

Reduction: Gain of electrons; decrease in oxidation number.

Reduction potential: A measure of the tendency of a given half-reaction to occur as a reduction.

Refining: Purification and treatment of a metal to give it properties that meet specific applications.

Representative element: An element in one of the A-groups in the periodic table.

Resonance: A concept in which the actual Lewis structure of a molecule or polyatomic ion is represented as a composite or average of two or more resonance structures.   The individual resonance structures themselves do not actually exist.

Resonance hybrid: The actual structure of a molecule or polyatomic ion that is represented by two or more resonance structures.

Reversible process: A process that occurs by an infinite number of steps during which the driving force for the change is just barely greater than the force that resists the change.

Ribose: A five-carbon sugar that is one of the building blocks of RNA.

RNA: Ribonucleic acid, a type of nucleic acid involved in protein synthesis (see transfer RNA and messenger RNA).

Roasting: Heating an ore in air, which converts sulfides to oxides.

Rust: Hydrated iron(III) oxide.

Rydberg equation: An empirical equation that allows the computation of the wavelengths of the lines in the emission spectrum of hydrogen.

Salt: Any ionic compound, except those containing OH^-.

Salt-bridge: A tube containing an electrolyte that connects the two half-cells of a galvanic cell.

Saponification: The base catalyzed hydrolysis of an ester.

Saturated: A solution that contains as much dissolved solute as it can hold in equilibrium with undissolved solute.

Saturated hydrocarbon: A hydrocarbon in which there are only C—C single bonds.

Scientific method: Observation, explanation, and the testing of an explanation by further observation.

Scientific notation: Numbers expressed as the product of a decimal number between 1 and 10 multiplied by 10 raised to an appropriate power, for example, $0.035 = 3.5 \times 10^{-2}$.

Second law of thermodynamics: Whenever a spontaneous event takes place, it is accompanied by an increase in the entropy of the universe.

Secondary alcohol: An alcohol with the formula

$$R-\underset{\underset{H}{|}}{\overset{\overset{R}{|}}{C}}-OH$$

Secondary structure: The coiling that takes place in a polypeptide chain.

Semiconductor: A substance that conducts electricity weakly.

Semipermeable membrane: A membrane that permits the passage of solvent molecules but not solute molecules.

Shell: All orbitals having the same value of n.

SI: Système International d'Unités; the International System of Units.

Sigma bond (σ-bond): A bond formed by the "head-to-head" overlap of two orbitals. The electron density is concentrated along the imaginary line joining the two nuclei.

Significant figures: Digits obtained in a measurement such that only the rightmost digit contains any uncertainty.

Silicone: A type of polymer in which the "backbone" consists of alternating silicon and oxygen atoms.

Silver oxide battery: A battery in which zinc serves as the anode and Ag_2O serves as the cathode in an alkaline electrolyte.

Simple cubic unit cell: A cubic unit cell with atoms, molecules, or ions only at the corners.

Simple sugar: See Monosaccharide.

Simplest formula: See Empirical formula.

Single bond: A covalent bond in which a single pair of electrons are shared between two atoms.

Slag: A relatively low-melting mixture of impurities that forms in the blast furnace and other furnaces used in refining metals.

Slaking: Treating lime (CaO) with water, which produces $Ca(OH)_2$.

Smectic liquid crystals: Rodlike molecules arranged in layers of parallel rods.

Soap: A solution of fatty acid anions.

Solubility: The amount of solute needed to give a saturated solution in a given amount of solvent.

Solubility product constant, K_{sp}: The equilibrium constant for the solubility of a salt. For a saturated solution, K_{sp} is equal to the product of the molar concentrations of the ions each raised to appropriate exponents.

Solute: A substance dissolved in a solvent.

Solution: A homogeneous mixture.

Solvation: The surrounding of a solute particle by molecules of the solvent.

Solvay process: Used to prepare Na_2CO_3 from $NaCl$, CO_2, and NH_3.

Solvent: The substance present in largest amount in a solution. In solutions containing water, water is almost always considered the solvent.

Solvent system acids and bases: The view that an acid is a substance that produces the solvent cation and a base produces the solvent anion. These cations and anions are the same as those produced in the autoionization of the solvent.

Specific activity: The number of emissions per second per gram of radioactive isotope.

Specific gravity: The ratio of the density of a substance to the density of water. It has no units.

Specific heat: The amount of energy needed to raise the temperature of one gram of a substance by one degree Celsius. Units are cal/g °C or J/g °C.

Spectator ion: An ion that does not participate in a particular chemical reaction.

Spectrochemical series: A listing of ligands in order of their ability to produce a large crystal field splitting.

Spin quantum number: s, which determines the direction in which the electron appears to be spinning. Its values are

$+\frac{1}{2}$ and $-\frac{1}{2}$.

Spontaneous change: A change that occurs by itself without outside assistance.

Stability constant: See Formation constant.

Standard atmosphere: A unit of pressure sufficient to support a column of mercury 760 mm high. It corresponds to exactly 101,325 Pa.

Standard cell potential: The potential of a galvanic cell at 25°C and 1 atm when all ionic concentrations are exactly 1 M.

Standard entropy, $S°$: The entropy that 1 mol of a substance has at 25°C and 1 atm.

Standard entropy change, $\Delta S°$:
$$\Delta S° = (\text{Sum of } S° \text{ of products}) - (\text{Sum of } S° \text{ of reactants})$$

Standard free energy change, $\Delta G°$: $\Delta G° = \Delta H° - T\Delta S°$. The magnitude and sign of $\Delta G°$ determine the size of the equilibrium constant for a reaction.
$$\Delta G° = (\text{Sum } \Delta G_f° \text{ of products}) - (\text{Sum } \Delta G_f° \text{ of reactants})$$

Standard free energy of formation, $\Delta G_f°$: The change in free energy when one mole of a compound in its standard state is formed from the elements in their standard states.

Standard heat of formation, $\Delta H_f°$: The enthalpy change associated with the formation of one mole of a substance in its standard state from its elements in their standard states.

Standard heat of reaction, $\Delta H°$:
$$\Delta H° = (\text{Sum of } \Delta H_f° \text{ of products}) - (\text{Sum of } \Delta H_f° \text{ of reactants})$$

Standard reduction potential: The reduction potential of a half-reaction at 25°C and 1 atm when all ion concentrations are 1 M.

Standard state: The natural state of a substance at 25°C and 1 atm pressure.

Standard Temperature and Pressure (STP): Reference conditions for problems involving gases; 0°C and 760 torr.

Standing wave: A wave whose peaks and nodes do not change position.

State: A particular set of conditions of pressure, temperature, volume, and number of moles of each component of a system.

State function: A quantity whose value depends only on the current state of the system and not on the system's prior history. The magnitude of the change in a state function depends only on initial and final states of the system and is independent of the path followed between these states.

State variable: See State function.

States of matter: Solid, liquid, or gas.

Stereoisomers: Isomers that have the same atoms bonded to each other but differ in the way their atoms are arranged in space.

Steroid: A lipid with a complex ring structure that possesses very high biological activity.

Stock system: System of nomenclature that uses Roman numerals to specify oxidation states.

STP: 0°C and 760 torr

Strong acid: An acid that is 100% dissociated in water.

Strong base: A base that is 100% dissociated in water.

Strong electrolyte: An electrolyte that is 100% dissociated in water.

Structural formula: A chemical formula that shows which atoms are bonded together in a molecule.

Structural isomers: Isomers that differ in the ways that the atoms are bonded to each other.

Sublimation: Conversion of a solid directly to a gas without passing through the liquid state.

Subshell: All orbitals of a given shell that have the same value of ℓ.

Substitution reaction: A reaction in which one atom replaces another in an organic molecule.

Substitutional solid solution: A solid solution in which an atom of the "solute" replaces an atom of the "solvent" in the lattice of the solvent.

Substrate: The substance acted on by an enzyme.

Supercooled liquid: A liquid at a temperature below its freezing point.   An amorphous solid.

Supercooling: Cooling a liquid to a temperature below its freezing point.

Supercritical fluid: A substance at a temperature above its critical temperature.

Superoxide: A compound containing the O_2^- ion.

Supersaturated: An unstable solution that contains more dissolved solute than it could hold if it were in contact with undissolved solute.

Surface tension: A measure of the amount of energy needed to expand the surface area of a liquid.

Surfactant: A substance that lowers the surface tension of a liquid and promotes wetting.

Surroundings: That which exists outside the system.

Synthesis gas: A mixture of CO and H_2.

System: That particular portion of the universe upon which we wish to focus our attention.

$t_{\frac{1}{2}}$:   See Half-life.

Temperature: A measure of the hotness of a body; an intensive quantity that measures the intensity of heat.

Termination step: In a chain reaction, a step in the mechanism that removes free radicals and thereby terminates the chain.

Tertiary alcohol: An alcohol with the structure

$$\begin{array}{c} R \\ | \\ R-C-OH \\ | \\ R \end{array}$$

Tertiary structure: The way a coiled polypeptide folds to give globular proteins.

Tetrahedron: A four-sided pyramid with triangular faces. A tetrahedral molecule has an atom in the center with other atoms joined to it which are located at the four vertices of the tetrahedron.

Theoretical yield: The maximum amount of product(s) that could be obtained from a particular mixture of reactants.

Theory: A tested explanation of the results obtained in many experiments.

Thermite reaction: The very exothermic reaction,
$$2Al + Fe_2O_3 \longrightarrow 2Fe + Al_2O_3.$$

Thermochemical equation: An equation whose coefficeints are interpreted as representing numbers of moles of reactants and products, and which is accompanied by the energy change for the reaction.

Thermodynamic equilibrium constant: An equilibrium constant computed from the equation, $\Delta G° = -RT \ln K$. For reactions involving gases, $K = K_p$.

Thermodynamics: The study of energy changes and the flow of energy from one substance to another.

Thio: In a chemical name, thio means sulfur in place of oxygen.

Third law of thermodynamics: For a pure crystalline substance, $S = 0$ at 0 K.

Tie line: A horizontal line on a boiling point diagram that connects the boiling point curve to the vapor composition curve.

Titrant: The solution that is dispensed from a buret.

Titration: An analytical procedure in which a solution, generally of known concentration, is added gradually from a buret to another solution where the solutes react. When the completion of the reaction is signaled by an indicator, the volume of solution added from the buret is carefully read and recorded.

Torr: A unit of pressure; 1 torr = 1 mm Hg.

Tracer study: The use of radioisotopes to follow the fate of certain atoms during chemical reactions.

Trans-: A term applied to a geometrical isomer in which two groups are located on opposite sides of some central point in the molecule or ion.

Transfer RNA (t-RNA): A small RNA unit that carries amino acids to their proper location along m-RNA during protein synthesis.

Transition elements: Elements located between Groups IIA and IIIA.

Transition metals: Transition elements.

Transition state: The brief moment during a reaction when the reactants have collided and are at the high point on the potential energy diagram for the reaction.

Triad: Any one of the three horizontal sets of elements in Group VIII (e.g., Fe, Co, and Ni).

Triglyceride: An ester of glycerol and three fatty acid molecules.

Trigonal bipyramid: A geometrical figure composed of two trigonal pyramids (pyramids with three-sided faces) sharing a common triangular face. A trigonal bipyramidal molecule has an atom in the center of this triangular plane and is joined to five other atoms which are located at the vertices of the trigonal bipyramid.

Triple bond: A covalent bond in which three pairs of electrons are shared between two atoms.

Triple point: The temperature and pressure at which the liquid, solid, and vapor states of a substance can coexist in equilibrium.

Triprotic acid: An acid that can furnish three H^+ per molecule.

Tritium: The radioactive isotope of hydrogen, ^3_1H, sometimes represented by the symbol, T.

Trivial name: The common name of a compound.

Uncertainty principle: There are limits in our ability to simultaneously measure a particle's speed and position.

Unit cell: The smallest portion of a crystal that can be repeated over and over in all directions to give the entire crystal lattice.

Universal gas constant: $R = 0.0821$ liter atm mol^{-1} K^{-1};
    $R = 1.987$ cal mol^{-1} K^{-1}; $R = 8.314$ J mol^{-1} K^{-1}.

Unsaturated: A solution capable of dissolving more solute.

Unsaturated hydrocarbon: A hydrocarbon having one or more
    carbon-carbon double or triple bonds.

Valence band: A band of energy levels in a crystal that is
    formed from the energy levels of the atom's valence shells.

Valence bond theory: Theory of covalent bonding that views a
    bond as being formed by the sharing of one pair of elec-
    trons between two overlapping atomic or hybrid orbitals.

Valence electrons: Electrons in the valence shell.

Valence shell: Shell with highest n that is occupied by electrons
    in an atom.

Valence Shell Electron Pair Repulsion theory: A theory used to
    predict molecular structure. It is based on the idea that
    electron pairs (bonding and lone pairs) in the valence shell
    of an atom stay as far apart as possible.

Van der Waals equation of state: A modified form of PV = nRT
    in which corrections for the finite volume of gas molecules
    and intermolecular attractions are applied to the measured
    volume and pressure of a real gas.

van't Hoff factor: The ratio of the measured ΔT_f to the ΔT_f
    calculated assuming the solute in a solution to be a nonelec-
    trolyte.

Vapor pressure: The pressure exerted by the vapor above a
    liquid. If the liquid and vapor are in equilibrium, this
    pressure is the equilibrium vapor pressure. Normally, the
    term vapor pressure is taken to mean the equilibrium vapor
    pressure.

Vapor pressure curve: The variation of equilibrium vapor pres-
    sure with temperature.

Vibrational frequency: The frequency with which atoms joined
    by a chemical bond vibrate back and forth, toward and away
    from each other.

Volt (V):   The SI unit of electrical potential or emf.
    1 V = 1 J/C.

Voltaic cell: See Galvanic cell.

Volume: The space occupied by a substance.

Volumetric analysis: Analytical procedures that make use of reactions in solution, where volumes and concentrations of solutions are carefully measured.

Volumetric flask: A special flask having a long thin neck that has a mark etched around it.   It is used to prepare solutions of accurately known molarity or normality.

VSEPR theory: See Valence Shell Electron Pair Repulsion theory.

Vulcanization: Treatment of natural rubber with sulfur, which forms sulfur bridges between adjacent polymer strands.

Wave function: A mathematical function (represented by the symbol, ψ, and obtained by solution of the wave equation) that describes the shape and size of an orbital, and which determines the energy of an electron in that orbital.

Wave mechanics: Theory of atomic structure based on the wave properties of matter.

Wavelength: The distance between successive peaks in a wave.

Weak acid: An acid that is less than 100% dissociated in water.

Weak base: A base that is less than 100% dissociated in water.

Weak electrolyte: An electrolyte that is less than 100% dissociated in water.

Weight: The force with which an object of a certain mass is attracted to the earth or some other object that it may be near.

Weight fraction: The number of grams of solute divided by the total number of grams of solution.

Weight percent (% w/w):   Weight percent = 100 x weight fraction

Wetting: The spreading of a liquid across a solid surface.

White phosphorus: A very reactive form of phosphorus consisting of P_4 molecules.

X-ray diffraction: The diffraction of X rays by the atoms in a crystalline solid.   A technique for studying the structures of crystalline solids.

<u>Zinc-carbon dry cell</u>:   The ordinary dry cell in which zinc serves as the anode and MnO_2 serves as the cathode.

<u>Zone refining</u>:   A method for producing very high purity solids. A thin cross section of a bar-shaped solid is melted and the molten zone moved from one end to the other.   Impurities collect in the molten zone.